A LANDSCAPE TRANSFORMED

A LANDSCAPE TRANSFORMED

The Ironmaking District of Salisbury, Connecticut

Robert B. Gordon

OXFORD

UNIVERSITY PRESS

2001

OXFORD

UNIVERSITY PRESS

Oxford   New York

Athens   Auckland   Bangkok   Bogotá   Buenos Aires   Calcutta
Cape Town   Chennai   Dar es Salaam   Delhi   Florence   Hong Kong   Istanbul
Karachi   Kuala Lumpur   Madrid   Melbourne   Mexico City   Mumbai
Nairobi   Paris   São Paulo   Shanghai   Singapore   Taipei   Tokyo   Toronto   Warsaw

And associated companies in
Berlin   Ibadan

Copyright © 2001 by Oxford University Press, Inc.

Published by Oxford University Press, Inc.
198 Madison Avenue, New York, New York 10016

Oxford is a registered trademark of Oxford University Press.

Library of Congress Cataloging-in-Publication Data
Gordon, Robert B. (Robert Boyd), 1929–
A landscape transformed : the ironmaking district
of Salisbury, Connecticut / Robert B. Gordon.
p.   cm.
Includes bibliographical references and index.
ISBN 0-19-512818-4
1. Iron-works—Connecticut—Salisbury Region—History.
2. Industrial Ecology—Connecticut—Salisbury Region. I. Title.
TN704.U52 C664   2000
338.4'7669141'097461—dc21   99-041825

1 3 5 7 9 8 6 4 2

Printed in the United States of America
on acid-free paper

Preface

The strong hand of the past rests heavily on today's environment. Past generations consumed resources, left wastes, changed habitats, and altered the shape of the land, sometimes irreversibly. They made choices about settlement patterns and transportation systems that endure for decades, centuries, or longer. The political institutions and policies they created, by determining the division of power between individual choice, community decision, and central authority, define the avenues we can use to address environmental issues. Inherited values long embedded in culture influence attitudes toward the environment as powerfully as modern preferences. While everyone must accept the limits on environmental change defined by natural science, past practices and established values are powerful determinants of current choices. We need history to understand our environment.

We also need analysis of the network of causes and effects that radiates from every choice made about the use of a natural resource. *Industrial ecology* is the new discipline that provides the tools to trace environmental consequences. When industrial ecologists look into the subject of this book, iron smelting, they see consumption of non-renewable ore and fuel, creation of solid wastes, and release of effluents into the atmosphere. With *life-cycle analysis* they discover how, until the iron eventually reverts to rust, its use makes further demands on the environment: if it is converted to steel and built into an automobile, petroleum will be burned, roads enlarged, and rubber tires worn out. Then, in a waste stream, additional energy will be used to recycle it, or landfill occupied used to store it.

This book examines the history and industrial ecology of a primary industry—ironmaking in Connecticut's Salisbury district—from its eighteenth-century inception through national prominence in the early Republic to its end in the twentieth century. It investigates the cultural context in which people made decisions about their use of technology and the environment. It examines the environmental consequences of a heavy industry that fully utilized a region's renewable energy resources and its non-renewable ore resources for nearly two centuries.

Where people in the Salisbury district dug ore, flushed away mine waste, built furnaces, clear-cut forest, dammed rivers, and piled up slag as they carried on a nationally important component of ironmaking, a visitor today finds a landscape of wooded hillsides, neat villages, and valley farms. Tourists come for concerts, art galleries, and bookstores; they go rafting or canoeing on the region's rivers. New residents arrive to escape from urban business and commerce. If these visitors or incomers happen to attend a lecture at a local historical society or browse in one of the district's libraries, they find evidence that the landscape now hosting leisure and the arts was once totally dedicated to heavy industry of the sort that we today would identify with polluted, dirty cities. A lake used by a summer camp may have been built to supply water to a forge. The landholdings of a forest preserve would have been assembled to grow fuel for a blast furnace. An attractive house in a rural community could have been an ironmaster's home; smaller dwellings now remodeled would have once housed artisans who worked in a mine or at a forge. The public library's book collection is likely to have begun in an ironmaster's home, and to be housed in a building erected by his heirs.

People in Salisbury made a gradual transition in the use of their land from extractive industry to residential services. They effected this transformation in their own way because they kept local control of their industrial enterprises. They valued economic diversity and independence. They created communities that placed high value on liberal education while cultivating respect for artisans' skills. Managers were often artisans, and artisans, managers. Nearly everyone involved themselves in some form of agriculture. While entrepreneurial, Salisbury people shunned those who placed acquisition of wealth and power beyond immediate needs. And they were aided by the particular topography and natural resources of their region.

The experience in Salisbury shows the powerful role of culture in shaping environment. It shows that, while understanding natural science is now an essential part of effecting thoughtful management of our environment, ultimately values and beliefs guide decisions about the natural world around us.

Chapter 1 presents a synopsis of some of the concepts of industrial ecology useful in the analysis of environmental change caused by Connecticut ironmaking. The next four chapters describe how European settlers and their descendants, acting as individual artisans, merchant capitalists, and artisan entrepreneurs, used the ore and energy resources of an otherwise unpromising region. They structured their ironmaking over 150 years to fill first a regional and then a national market niche. Chapter 6 explores the cultural and geographic context in which the Salisbury ironmakers made their decisions about technology and resource use through the first half of the nineteenth century. Chapters 7 and 8 show how, confronted with the need to adopt new technology as the nineteenth century progressed, the established Salisbury families turned to other uses of the region's resources, leaving ironmaking to incomers who held different values. Chapter 9 shows the region's transition from dependence on extractive industry to new uses of the land and its resources. Chapter 10 offers an interpretation of the district's industrial ecology. Appendix 1 catalogs the region's ironworks. Appendix 2 contains a brief description of the ironmaking techniques used in Salisbury.[1] The varied units of measure used in the iron industry complicate interpretation of production data. They are explained in appendix 3. A separate book deals with the industrial archaeology of the region.[2]

Acknowledgments

This book could not have been written without the help of Frances Kemmish, who searched the region's historical societies, libraries, and town archives for primary sources. Frances's research has taken her to graveyards in search of birth and death dates on headstones, and to land records to trace the ownership of houses occupied by artisans, managers, and capitalists. She tramped through woods and brush in search of long-forgotten forge and furnace sites.

Much of the history of Salisbury ironmaking resides in records and photographs held at local historical societies. I thank Doris Longaven, Marion Stock, Ronald Jones, and Gabriel Seymour of the Falls Village–Canaan Historical Society; Michael Gannett and Maureen Prentice of the Cornwall Historical Society; Laura Riva of the Salisbury Association; Emily Hopson of the Kent Historical Society; Nancy Beveridge of the Litchfield Historical Society; and Ginny Moskowitz, Salisbury's town historian, for their assistance in finding documents and photographs.

The Holley family left over 9,000 letters and other documents now in collections at the Connecticut Historical Society and the Salisbury Association. Byron Scott and his staff of volunteer readers have undertaken the task of abstracting and transcribing the letters in Salisbury. I thank Byron for making this material available for my research.

Edward Kirby generously shared the research on the iron industry of the northwest undertaken for his book *Echoes of Iron*. Fred Chesson, Fred Hall, Matthew Kierstead, Gregory Galer, Rovert Grzywacz, and Walter Landgraf generously shared the results of their research on northwestern Connecticut's industrial history and archaeology with me. Dr. William Adam allowed me to examine his extensive collection of family papers relating to the Forbes and Adam ironworks. Fred Chesson allowed me to use the collection of letters and photographs assembled by the late Charles Rufus Harte now in his care. Anne Knowles generously shared her insights on the geography of nineteenth-century American ironmaking with me.

At Yale, Margit Kaye and Frederick Musto helped me find maps in the collections of the Sterling Memorial Library. Brian Skinner explained the origin of the Salisbury ores and the region's bedrock geology. William

Sacco at the Peabody Museum spplied his skills to recording Salisbury scenes and reproducing old photographs.

Conversations with Carolyn Cooper, Patrick Malone, and Michael Raber, my colleagues in numerous industrial archaeology projects, have contributed in many ways to my research in northwestern Connecticut.

Research for this book was supported by grant 9631262 from the National Science Foundation.

Contents

A LANDSCAPE TRANSFORMED

ONE

Industrial Ecology in
Historical Perspective

Industry consumes natural resources and makes wastes as it manufactures and delivers products to consumers. Subsequent use of a product—its eventual discard, recycling, or storage in a waste depository—puts additional demands on the environment. Decisions made by many different individuals direct the progress of a product through manufacture, use, and disposal. In the past, each decision maker along this chain responded to concerns that encompassed only a fraction of the product's progression from raw materials to ultimate fate. No one had much reason to enlarge these decision horizons as long as natural resources remained abundant, and the industrial impact on the environment was small compared with natural processes of environmental change.

Now people in the western industrialized nations realize that their consumption of goods and services could change the environment in ways that rival natural causes. Their heightened awareness led scientists and engineers to start systematically investigating the life cycles of industrial products. These investigators soon found that Western industry has created a web of resource use so complex that tracing the demands made by even a single, simple product on the environment requires the new analytical methods of industrial ecology.[1] Industrial ecologists see the farm and the factory as the main sources of environmental change caused by people. With their focus on the factory, along with its associated mines, power plants, and transportation systems, they search out the consequences of consuming natural resources to make and use material goods and generate energy. They study resources consumed, wastes released, and the fate of discarded products. They may include in their research advocacy of, and searching for, means to minimize the environmental impacts of industry. They look to a future where recycling eliminates all wastes and where energy comes from renewable resources. Most see a guiding principle in sustainability, the concept that each generation should leave to the next undiminished opportunities for fulfillment of material needs.

People make their decisions about the production, consumption, and disposal of material goods in terms of the costs and benefits they perceive, and they may be unwilling to bear extra costs for environmental benefits that offer them no immediate rewards. The values signaled by markets and studied by economists are determined by culture, often through behavior so common and widely accepted as to be rarely examined or questioned. Values deeply embedded in society, such as minimizing consumers' costs and the premium governments and businesses place on growth, collide head-on with the steps toward a sustainable economy envisioned by industrial ecologists. Difficult as are the problems of making accurate life-cycle analyses, they pale before those of understanding or changing the values implicitly accepted in our culture that stand at odds with sustainability. An ecology of industry has to deal with the culture and the values held by people, as well as with technical analyses of materials and energy flows. History sheds light on the long-term consequences of industrialization and on how a community's values affect decisions its citizens make about the environment.

Natural resources are the starting point for analysis in industrial ecology. Pessimists fear that industries producing primary materials may exhaust our natural resources, and everyone knows that resource extraction alters the environment and creates wastes such as mine spoil and mill tailings. Extraction, exhaustion, and waste production are physical components of the industrial ecology of supplying primary materials to manufacturers.

Natural processes create new material and energy resources all the time. We call these resources *renewable* if we use them no faster than their replacement rates. Managing a forest so that the wood harvested each year equals the amount of new wood formed by tree growth makes a renewable resource that can supply materials (timber or other wood products) or energy (wood fuel or charcoal) indefinitely. Water drawn from a stream at a rate not exceeding its natural flow, passed through a waterwheel or turbine, and returned to the stream is a renewable energy resource.

Ongoing processes form petroleum, mineral coal, and ores too slowly for them to be renewable resources. When we use these materials, we draw down stocks created in the distant past. Once mined, these non-renewable resources are gone forever. Abundance determines the future availability of the non-renewable resources. Minerals containing iron, aluminum, magnesium, and silicon make up a large fraction of the earth's crust; they will always be available when needed. People may mine out conveniently placed, easily accessible, readily processed, rich deposits of the abundant minerals but will always be able to find more. However, geologists foresee possible worldwide exhaustion of resources such as helium, copper, zinc, lead, tin, chromium, tungsten, silver, or gold, materials so scarce in the earth's crust that they can be mined only where natural enrichment processes have concentrated them into ores.[2]

Resource exhaustion is the easiest issue to deal with in the industrial ecology of primary materials. Economists find no evidence that consumption of mineral resources has as yet ever slowed the economic growth of the industrial nations. Prospectors located new mines, metal-lurgists improved recovery from lower grade ores, and engineers found satisfactory substitutes for any material not readily available from an established source. These successes lead some people to believe that the industrial nations can deal equally well with any future resource scarcity.

The resource optimists overlook the environmental consequences arising from ever larger scale mining operations. Some of the environmental and social costs of getting non-renewable resources—abandoned pits, waste piles, and mine shafts—are familiar to residents of old mining towns. The chemical effects of mining, often less immediately visible, can be more deleterious. Many of the scarcer metals are found as sulfur-bearing minerals that react with water entering mines to form dilute, but nevertheless destructive, sulfuric acid that often flows into rivers or lakes. Smelting these minerals releases sulfur dioxide that, unless trapped, makes its way into the atmosphere.

Although exhaustion is not a concern, using renewable resources has environmental costs. Renewable energy from waterpower requires dams that create lakes out of previously free-flowing rivers, block fish migrations, and trap sediment that previously enriched downstream floodplains. A working forest, one managed for continuous wood production, will have trees in successive stages of growth, none of them mature, and tree species that differ from those in a forest disturbed only by natural events. The working forest will not provide services such as undisturbed wildlife habitat, stately trees, or wilderness experience.

Industrial ecologists must also deal with wastes. "Consumption" of materials and fuels means transforming them to other, less-useful forms that people discard or recycle. Recycling has a long history: prehistoric white smiths routinely recycled bronze artifacts. Nineteenth-century English ironmasters resmelted much of the slag left behind in the British Isles by Roman ironmaking. Michigan copper-mining companies built dredges to recover and rework mill tailings earlier dumped in Torch Lake. Scrap dealers recycle massive steel structures such as ships or railcars and machinery such as lathes, engines, or electrical generating equipment withdrawn from service. Despite these efforts at recycling, industrialists in the industrial nations have not matched their successes in managing material and fuel resources with corresponding accomplishments in handling wastes. Managers, consumers, and regulators have dealt with waste as a problem separate from production, searching for the cheapest possible mitigation methods, usually "end-of-the-pipe" techniques where wastes are processed as the last stage of manufacturing or immediately before release from an exhaust pipe.

Industrial ecologists offer two solutions to the waste problem: they would organize factories so that each uses another's waste as its raw material. In *design for the environment* they seek processes and designs that minimize creation of wastes in the making and use of a product. These techniques can reduce future production of industrial wastes. However, massive amounts of potential useful materials have accumulated in industrial middens, landfills, and other repositories. Some of these are environmental hazards. Some, once considered waste, may be valuable now: smelter tailings decades or even centuries old may be richer than the grades of virgin ores now available for mining.

Natural resources and wastes are obvious components of any region's industrial ecology. Were these the only components, the work of industrial ecologists would be much simplified. They could make predictions of environmental impact on the basis of topographic, geological, and engineering data alone. Engineers face formidable problems in completing life-cycle analyses, in design for the environment, and in finding ways to advance sustainable industrial production and consumption. However, they can work toward solutions with their established principles and practice. When they have to deal with the values and preferences embedded in Western society that determine how people allocate resources for their material goods and services, engineers and industrial ecologists enter unfamiliar territory where they lack the guidance of established method and theory.

People do change the values embedded in their culture, sometimes quite rapidly after intervals of stasis. Through the colonial years and into the early nineteenth century, New Englanders were willing to forgo a seventh of each week's productive labor and make difficult journeys to spend their Sundays in unheated halls listening to their ministers' sermons. They used much of the wealth they created to erect the architecturally distinguished meeting houses that grace village greens today. As recently as forty years ago, few people in a New England town would conduct business or trade on Sundays. Even affluent Yankee householders saved bits of string and fragments of soap for remelting, and generally made do with modest material means. Then, within a few decades, the grandchildren of these thrifty parents adopted distinctly different values that place powerful new demands on the environment.

Decisions about energy consumption, travel, packaging, and other aspects of material life that impinge on the environment are as culture bound as those determining how to spend Sundays. Although the evidence for change in values is clear, scholars find it difficult to discover how these changes are effected. Industrial ecologists can explain consequences and offer choices. Attainment of their goals for a sustainable economy is bound up with popular values that, if they change, do so for reasons difficult to discern.

History offers the perspective of decades or centuries on cultural change and reveals how shifting cultural values direct peoples' impact on the environment. We can improve our chances of separating and identifying the physical and cultural components of a region's industrial ecology by studying a community whose economy placed heavy demands on its natural resources to support a single, dominant industry. If the industry remained active for a century or more, we can follow the long-term environmental changes it caused. Ironmaking in Connecticut's Salisbury district (figs. 1.1, 1.2, 1.3) meets these conditions, offering us the opportunity

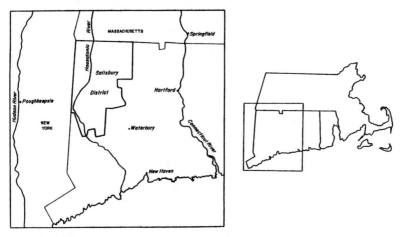

Fig. 1.1. *The Salisbury ironmaking district occupied the northwestern corner of Connecticut, where it was traversed by the Housatonic River.*

to study the industrial ecology of an extractive industry in the perspective of history.[3]

Even before the colonial government opened Connecticut's undivided Western Lands to settlement in 1737, adventurers had found iron ore deposits unmatched in New England. Newly arrived settlers discovered the forest and water resources they needed for iron smelting, while the growing New England economy created new demand for iron products. Aided by colonial investors, they opened mines, built forges and furnaces on the region's abundant streams, and coaled wood from its forests. They carried ironmaking into the surrounding towns and built a high reputation for their metal. Customers who wanted iron of the best quality soon began to demand Salisbury iron in preference to that made elsewhere.

In 1776 patriot artisans used Salisbury's mines, furnaces, and forges to make armaments for the Continental army. In the first decades of the early Republic, they supplied parts for steamboat engines and the castings, forgings, and raw material that ingenious Yankees used to launch New England's new manufacturing industries. Thus, Salisbury district forges supplied Eli Whitney with both the tools and the metal stock he needed to initiate manufacturing with water-powered machinery and division of labor. In the following decades, district artisans forged axles and locomotive tires for the nation's new railways. For over fifty years they were primary suppliers of iron to the national armories. They avoided use of mineral coal, instead assuring themselves of a sustained fuel supply by acquiring woodland that they then managed for continuous yield of wood that could be converted to charcoal. They saw their ore as abundant without bound and in fact never came close to exhausting it.

Salisbury artisans invented and manufactured explosive shells and large cannon for the federal government during the Civil War. After the war, they supplied the Union Pacific and other major railroads with wheels that

Fig. 1.2. Sixteen towns made up the Salisbury district. The new towns of North Canaan, Warren, and Morris (formerly the South Farms parish) were set off from their original, parent towns, as shown by the dashed lines.

earned a national reputation for their durability. Then, beginning in the 1870s, a new generation of district ironmasters, comfortable with their success and with other interests drawing their attention, ignored the new ironmaking techniques that their competitors in other districts adopted. They traded on their long-established reputation and let their industry drift into obsolescence. Sales gradually declined, and market forces finally extinguish Salisbury ironmaking in 1923, after nearly 200 years of production.

Other regions of the United States that, like the Salisbury district, once supplied primary industrial materials and fuels now face unresolved environmental and social consequences arising from their now defunct mines and smelters. The anthracite mining district of Pennsylvania is scarred by culm piles and mine fires, and its people still live in the shadow of the social conflicts that accompanied mining. In northern Michigan and western Montana, remains of metal mining and smelting dominate the landscape. Here the people of once-large communities are searching for ways

Fig. 1.3. Principal villages associated with ironmaking within the towns of the Salisbury district.

to use their mining and metallurgical heritage to sustain themselves while simultaneously dealing with the environmental consequences of their industries.[4]

Three-quarters of a century have passed since iron mining and smelting ended in Salisbury. Many North American mining districts have been defunct as long, while those in Pennsylvania, Michigan's Keweenaw, and at Montana's Butte are well into a half century of idleness. Few of these regions have made the progress toward the transformation of the landscape to new uses realized in the Salisbury region. While accidents of geography account for some of this difference, aspects of Salisbury's culture and the values held by its people shaped its particular industrial ecology.

This study of the Salisbury district will look for evidence on the environmental and social consequences of ironmaking. Salisbury residents left us thousands of letters, along with diaries, account books, and town meeting minutes. Material evidence supplements these records. Additionally, surviving artifacts reveal unrecorded aspects of work at the mine, forge, and furnace. Alterations of the landscape show us the changes wrought by those who actually exploited the region's natural resources, making an additional record of past peoples' attitudes toward their environment. These data can demonstrate how the consumption of non-renewable resources affected the region's industry and environment, and the conditions under which production of a primary material was sustained with renewable fuel and power resources. Documents record who made the decisions that directed the course of the industry, and the role of the region's culture in shaping the attitudes of the investors, managers, and artisans who converted local natural resources into products sold on a national market.

TWO

Resources Discovered

The adventurers who entered Connecticut's Western Lands in 1730 began ironmaking more than a hundred years after colonists first exploited the ore and fuel resources of British North America. The early colonists who set about making iron for export met with ill fortune: in 1621 Indians massacred the artisans who had just completed a furnace and forge at Falling Creek, Virginia. Scarce capital, inadequate skills, and poor transatlantic communication bankrupted the proprietors of the Saugus, Massachusetts, and New Haven, Connecticut, ironworks by 1675.

When King George I got Parliament to restrain trade between England and Sweden in 1717, British manufacturers, cut off from their supplies of Scandinavian iron, began investing in American forges and furnaces. Conclusion of the seventeenth-century Indian wars had left large areas rich in timber and ore along the east coast safe for industry. New immigrants, primarily from Britain and Germany, brought their metallurgical skills to America, and colonists supported by British investors built ironworks first in Maryland and then in Pennsylvania, Virginia, and New Jersey, to produce metal for the export market. Americans in the Middle Atlantic colonies made enough iron by 1750 to provoke British regulation of their trade. The colonists made themselves the world's third-largest iron producers by 1775 and, despite the predominance of agriculture, had firmly established industry in British North America.[1]

New Englanders lagged behind the Middle Atlantic colonists in ironmaking. Artisans from the failed Saugus works in Massachusetts slowly reestablished smelting on a small scale and by 1730 were building new works in the southeastern part of their colony. In New York, Robert Livingston had by 1685 gained control of an enormous manor adjacent to northwestern Connecticut. In 1730 he wanted to add iron to his manor's products so that he could ship metal down the Hudson River to colonial and overseas customers. However, neither Livingston nor the Massachusetts ironmakers had anything like the high-grade ore resources discovered by the adventurers in Connecticut's Western Lands.

Fifty-two years after English colonists established themselves in Connecticut, James II sent Edmund Andros to British North America to set up a unified government over the New England colonies. In a series of moves that culminated in the famous charter oak incident, the Connecticut General Assembly sought to protect the colony's unsurveyed, undivided, and largely unknown western wilderness by granting title temporarily to the towns of Hartford and Windsor. Then, with Andros's claims repulsed and the colonial charter secure, the assembly attempted to regain what was by then known as the Western Lands. The citizens of Hartford and Windsor found it convenient to forget that the assembly intended temporary grants. A prolonged dispute between the assembly and the towns, marked by some violence, ended in a 1724 compromise: land west of the Housatonic River that later became the towns of Salisbury and Sharon and the land to the east, later to be Goshen, Cornwall, Canaan, Kent, and Norfolk, was returned to the colonial government.[2] In the meantime adventurers from Hartford, Windsor, and Lebanon had purchased land for the town of Litchfield, where, by 1720, they had started building a new community.[3] Farther west, several Dutch families purchased land they thought was in New York or moved into what turned out to be Kent and Salisbury when Connecticut and New York settled their boundary in 1731. Livingston's New York manorial lands also extended into an area that the 1731 boundary agreement placed in Connecticut.

While the Dutch and Livingston incursions from New York were under way, adventurers from Massachusetts also saw opportunity for profit in the northwestern Connecticut wilderness. After passing through a depression in the late seventeenth century, Connecticut's colonial economy improved in the 1730s, aided by the opening of new land and population growth. Connecticut was changing from an internal, self-sufficient economy to one producing agricultural products for outside markets.[4] Thomas Lamb (from Springfield), Ezekiel Ashley, and John Pell (both from Sheffield) secured several land grants and water privileges from Connecticut's colonial government. Ashley and Pell used their skills as professional surveyors to locate valuable natural resources for themselves, including Ore Hill, which would prove to be the region's largest iron deposit. Daniel Bissell in Salisbury held Ore Hill on unsurveyed land under an old colonial grant. Ashley and Pell purchased the grant from Bissell and in 1734 got the General Assembly to authorize both a survey to clear their title and a corporation to open a mine.[5] Meanwhile, Lamb bought up land grants that earlier adventurers had gotten through deals with Indians to gain control of the iron ore bed later known as the Davis mine. With his acute eye for valuable property, he acquired water privileges on Salisbury's best streams.[6] Another group of adventurers located a large ore bed before 1732 in an area that would later be part of the town of Kent.[7]

Connecticut's government finally opened the northwestern corner of the colony to settlement by dividing its Western Lands into the townships of Goshen, Cornwall, Canaan, Kent, Norfolk, Salisbury, and Sharon. Between 1737 and 1738 it sold each town at auction to a company of proprietors, many of whom moved onto their newly acquired land. The farmers among the new settlers planted grain in the limestone valley that ran through Sharon into Salisbury and cleared level fields on old glacial lake bottoms along the Housatonic and Blackberry Rivers. However, the new proprietors found that the upland forest on steep hillsides that covered much of their new domain held little promise for farming except as pasture. When they sought alternative economic opportunity, they found ironmaking attractive.[8]

Previously, all New Englanders had used bog ore at their ironworks. The deposits were small, and the ore likely to be rich in phosphorus. The large beds of limonite in northwestern Connecticut offered the potential for development of an iron industry unmatched elsewhere in New England. To make iron from this ore, the new arrivals in the northwest needed charcoal fuel and a source of power to drive the bellows of the hearth and the massive helve hammer of a bloomery forge. The region had an abundance of land suited for growing wood that could be converted into charcoal. It also had numerous streams with a relatively uniform flow throughout the year, and many places on them (called *water privileges*) where millwrights could build waterpower systems at relatively low cost. Anyone who wanted to build a forge or furnace in the former Western Lands could find a site to develop with a modest investment.

To raise capital to get mining and smelting operations under way, the adventurers who had located the ore beds sold shares to downstate investors. Ashley and Pell got Jared Eliot of Killingworth, John Eliot of New Haven, Robert Walker Jr. of Stratford, Elisha Williams of New Haven, Martin Kellogg of Wethersfield, and Richard Seymour of Hartford, as well as Philip Livingston, proprietor of the manor in neighboring New York, to invest.[9] The discoverers of the Kent mine brought in Alexander Wolcott of New Haven, Robert Walker, Elisha Williams, Jabez Hurd of Newtown, Jared Eliot, and David Lewis of Stratford as proprietors. These investors were prominent in colonial affairs: Elisha Williams was a minister, the rector of Yale College from 1726 to 1739, and a colonial politician; Robert Walker Jr. was a lawyer;[10] and Jared Eliot was a minister, physician, advocate of scientific agriculture, and, in 1762, recipient of a gold medal from London's Society for the Encouragement of the Arts for making iron from black sea sand.[11] (The black sand contained fine grains of magnetite that he reduced in his son's bloomery forge in Killingworth.) These investors provided the capital needed to get mining and smelting started, and they did not have to assume any active responsibilities for the management of the properties.

Although the first settlers in the Western Lands found forest everywhere, they did not always find a lot of mature timber. Indians in Salisbury had burned over much of the woodland every fall to destroy old grass and vegetation, and to ring deer for slaughter. Substantial trees survived only in the wetlands. Litchfield's settlers in 1718 found that fires Indians set for their hunting had nearly cleared many of the neighboring hills of trees. Kent settlers likewise discovered little wood in town because the Indians regularly burned the country. Goshen, however, had chestnut trees ranging to more than seven feet in diameter, as well as abundant chestnut, oak, and hemlock.[12]

Colliers could make excellent charcoal from hardwoods, which grew well throughout the Western Lands. After woodsmen cut over an area, new trees big enough for coaling grew back in twenty to thirty years (fig. 2.1). Because of the cessation of Indian burning in Salisbury, people there had more wood in 1803 than they had in 1740, despite their clearance of woodland for fields and fuel.[13] The forest offered the possibility of a sustained fuel supply for ironmaking, if properly managed. Wood could be cut for coaling almost anywhere in the region.

The Salisbury iron ore beds lay along the boundary between the Stockbridge dolomitic limestone and the Berkshire schist because weathering of siderite (iron carbonate) in the limestone formed the ore, which was then retained in pockets in the schist (fig. 2.2).[14] Once miners started digging into the ore beds, they found that the main ore bodies, because of their formation as a weathering product and subsequent glaciation, were beneath as much as seventy-five feet of glacial till and hardpan (a mixture of clay and pebbles). Additionally, none of the beds was near a water privilege that could power a smelting works. Ore would have to be hauled a considerable distance to a smelting site.

In the district's first decades, only public paths led from the mines to sites that had the waterpower needed for smelting. People considered travel outside their own towns hazardous.[15] In 1736 Thomas Lamb loaded ore in saddlebags for a trip of 3.5 miles over trails from his mine in Lakeville to his bloomery forge at Lime Rock (figs. 1.3, 2.3). Shallows and falls made navigation of the Housatonic virtually impossible. Hence, to build up significant iron production and reach customers elsewhere in the

Fig. 2.1. *This coppiced forest in Cornwall has been cut and the wood stacked ready for the colliers who will convert it to charcoal. Uncut trees stand in the background, while young trees show through the snow on the distant hills. (Courtesy of the Cornwall Historical Society.)*

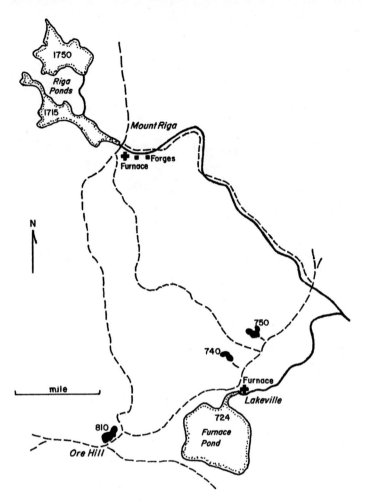

Fig. 2.2. A) Map showing the principal iron mines in the town of Salisbury, the two blast furnaces operating in 1810, and the two finery forges on Mount Riga. Elevations are shown in feet. Getting ore to the Mount Riga furnace required a climb of nearly a thousand feet up mountain roads now abandoned as too steep to use or maintain.

colony, Salisbury district towns had to build roads. The colonial government first addressed inland transportation with its 1643 act requiring each town to appoint two surveyors of roads, who were to call out every able-bodied man between sixteen and sixty to make and mend roads on appointed days. It added another law in 1679 designating roads between towns as king's highways. Since the legislature appropriated no money for these projects, the towns accomplished little.[16] Thus, the Greenwoods Road, completed in 1764 and running diagonally across the northwestern corner of Connecticut, was reported in 1766 to be "in great want of amendment."[17]

Taxation for whatever purpose, including roads, was a sensitive subject

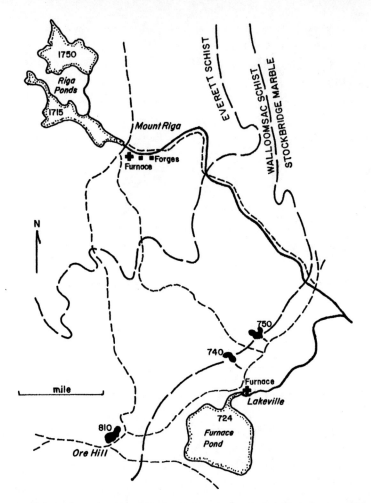

Fig. 2.2. B) Boundaries between the principal types of rock near the iron mines. The erosion-resistant Everett schist (northwest of the long-dash line) forms the highest ground, and the Stockbridge marble (southeast of the long-dash line), the lowest. The ore originated in the marble and was retained in pockets in the schist as the marble weathered away.

in the early Republic. Townspeople were little inclined to either pay taxes or contribute involuntary labor to make roads that would primarily bene-fit others. They felt that, as a matter of principle, beneficiaries should pay for the cost of services received. The government in Hartford responded to this attitude by privatization: it began chartering turnpike companies in 1792. Turnpike stock quickly became a popular investment for people who had acquired capital through the resumption of trade after the war. Mer-chants, artisans, and farmers usually financed the roads through their own towns, thereby making each road a locally controlled enterprise function-ing as a private utility under state regulation.[18]

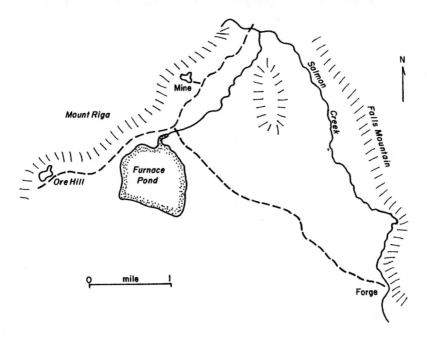

Fig. 2.3. In 1736 Thomas Lamb carried ore from the bed he owned northeast of Furnace Pond over trails to his forge at Lime Rock, a distance of 3.5 miles. Because the trail crossed the valley (low because it is floored by marble) between Mount Riga and Falls Mountain, the horses had a climb of only 300 feet up the trail. They then went down 445 feet to the forge site.

The turnpike companies either improved existing roads or, in some cases, built entirely new ones. Travelers and shippers of goods soon noticed a dramatic improvement in overland travel. Benjamin Silliman reported in 1819 that "the fine turnpike on which we commenced our journey, was, but a few years since, a most rugged uncomfortable road; now we passed it with ease and rapidity."[19] The Greenwoods Turnpike, completed in 1799 from New Hartford to Canaan, was the first to reach the Salisbury district. The Canaan and Litchfield followed the next year, and a dozen others were completed in the next decade (fig. 2.4).

The turnpike roads made it much easier for forge proprietors to fetch ore from the mines. Forges in Kent, Washington, and Litchfield got most of their ore from the Kent mine (fig. 2.5). The route to the Morgan forge followed town roads and required a net climb of 440 feet over the 4.4-mile trip. However, an 0.8-mile descent of 124 feet from the mine brought a teamster to turnpikes that reached the forges in New Preston, Woodville, and Bantam. These turnpikes afforded better surfaces and easier gradients than the town roads.

Roads had to cross the numerous streams that gave the district its abundant waterpower resources. Carpenters built timber trestles constructed like a barn floor with girts and planks to place between stone abutments

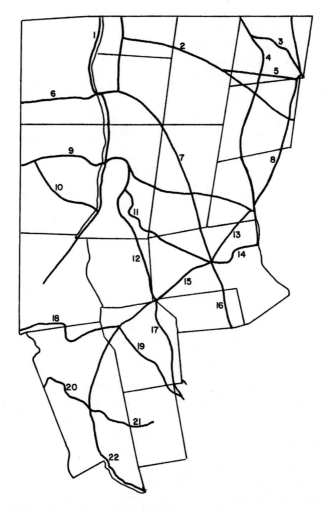

Fig. 2.4. Turnpike companies began building roads through the northwest in 1797 and within a decade had completed a network that connected the region's ironworks to the Hudson River ports and the principal cities and towns of Connecticut. The complete network included these turnpikes: 1, Warren; 2, Greenwoods; 3, Sandy Brook; 4, Waterbury River; 5, Hartland; 6, Salisbury & Canaan; 7, Canaan & Litchfield; 8, Still River; 9, Goshen & Sharon; 10, Sharon & Cornwall; 11, Litchfield, Goshen & Cornwall; 12, Cornwall & Washington; 13, Torrington; 14, Litchfield & Farmington; 15, New Milford & Litchfield; 16, Straits; 17, Washington; 18, New Preston; 19, New Preston & Washington; 20, New Milford & Sherman; 21, New Milford & Roxbury; 22, Ousatonic.

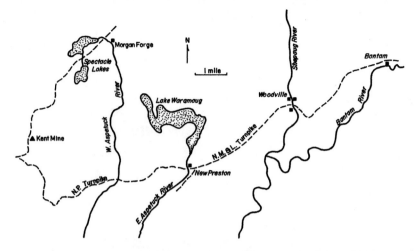

Fig. 2.5. Map of the roads from the Kent mine to bloomery forges (indicated by square symbols) that depended on this mine for their ore. Town roads ran from the mine to the Morgan forge and to the New Preston Turnpike. At New Preston teamsters transferred to the New Milford & Litchfield Turnpike.

for most bridges. They placed piers to carry a series of trestles across the larger streams. The Housatonic was the most difficult to cross. Burrall's Bridge, erected in 1760 just below the Great Falls, first offered Salisbury and Canaan residents a way over, rather than through, the river. To the north, Dutcher's Bridge formed an important link in the road used to carry ore from Salisbury to the ironworks in Canaan.[20]

THREE

Independent Artisans

By 1730 New England colonists needed increasingly large amounts of iron for their expanding economy. Shipsmiths forged iron fastenings used in the vessels they built for the coastal and West Indian trades. The mariners who sailed these ships wanted large, strong iron anchors. Millwrights needed waterwheel axles and gudgeons, spindles, and numerous other iron components for gristmills, sawmills, fulling mills, and oil mills. Builders of the forges and furnaces that smelted and shaped iron products had to have iron hammerheads and forge plates. The pioneers on the frontier in New York and northern New England wanted massive iron kettles for boiling potash, usually the first cash crop they got off their newly cleared land. Everyone needed nails.

Building a bloomery forge offered an adventurer in Connecticut's Western Lands the easiest way to start making iron. One man could run a forge, although a helper made the work easier. The bloomery proprietor needed less capital than would be required for other types of ironworks. The region had plenty of easily developed water privileges of the right size to power a bloomery forge. Although it took skill and practice to make high-quality metal, a forge owner or hired hand could learn enough of the bloom smelting technique from an experienced smith within a few months to make serviceable metal. Iron of ordinary quality satisfied most people's needs in the early days of the northwest. If the weather were bad, ore or fuel were unavailable, crops demanded attention, or the market of iron were slow, the proprietor could easily shut down his forge at short notice and restart it as soon as conditions improved.

Although a bloomery forge could be part of an enterprise employing fifty or more hands, it could also be little more than a smithy in size and complexity. A farmer could accumulate enough money to build one. Alternatively, a number of individuals might take shares in a forge run by a single artisan. The proprietors of a mercantile business, or of grist or sawmills on the same or a nearby water privilege, could easily add a bloomery to their other enterprises.

To deal with the varied aspects of forge operation, an entrepreneur in ironmaking needed several particular skills in addition to smelting technique: the ability to locate and assemble natural resources, raise capital, build the forge equipment, and market the product. The talents of the entrepreneurs who initiated ironmaking included different proportions of the speculator, capitalist, manager, and artisan. Some hopeful ironmakers lacked one or more of these essential skills, failed in their endeavor, and soon sold out to others.

Martin Kellogg, one of the earliest adventurers in the northwestern wilderness, initiated ironmaking near Litchfield. Indians had carried young Kellogg from his Deerfield home and into captivity in 1704. He escaped within a year and, after settling in Wethersfield in 1716, began real estate speculations in the Western Lands. Kellogg's 1729 purchase of a hundred acres abutting the Bantam River west of Litchfield's recently built sawmill gave him a good site for a bloomery forge. Here the falls of the Bantam offered a solid foundation for a dam, and the flow of the river could give him adequate power. Level land nearby would accommodate the forge building. Kellogg planned to smelt ore brought over trails from the mine in South Kent. He persuaded prominent men in the long-settled parts of the colony who were also investing in the northwest's iron mines to join him in financing the forge. Elisha Williams, Jared Eliot, and Robert Walker Jr., proprietors of the mines at Ore Hill and Kent, bought shares in the Bantam ironworks.[1]

Thomas Lamb, a land speculator from Massachusetts, built the first forge in the town of Salisbury. He bought the water privilege (fig. 3.1) at the falls on Salmon Creek at Lime Rock in 1734, moved his family in from Springfield, and by 1736 had the forge built. He hired two bloomers, Ben-

Fig. 3.1. Thomas Lamb chose this site on Salmon Creek in Lime Rock for his bloomery forge because the falls offered abundant power to run his forge hammer. Holley & Coffing converted the bloomery to a finery forge about 1825, and in 1830 built a blast furnace to supply the pig to the forge. Both operated until 1857. The forge and furnace stood on the right bank, where the furnace hearth is still in place.

jamin and Thomas Austin of Suffield, to smelt the ore brought in saddle-bags from his mine near Lakeville (see chap. 2). The Austins prospered with ironmaking, and Thomas became a prominent member of the Salisbury community.[2] Philip Livingston, proprietor on the manorial lands and a forge just over the border in New York, saw here an opportunity to enlarge his ironmaking enterprise: he bought a three-fourth interest in Lamb's Lime Rock forge in 1740 and gained full ownership by 1750.[3]

Miners, bloomers, their families, and their draft animals had to be fed and housed. Everyone in an ironmaking community needed gristmill and sawmill services, and a water privilege that powered a forge could also run these mills. Lamb soon had both a sawmill and a gristmill running near his Salmon Creek forge. He extended his operations in 1739 when the Salisbury proprietors voted him the right to build a sawmill on Fell Kill (Riga Brook); he soon added a gristmill there.[4]

Many ironworks proprietors in the northwest continued combined operations like these well into the nineteenth century. Jacob Bacon and Daniel Parke of Salisbury built a forge and gristmill in 1748 on Succonups Brook, the outlet of the pond at Chapinsville. Hezekiah Camp, who had moved to Salisbury from New Haven in 1746, acquired it in 1759. Hezekiah passed the forge on to his son-in-law Phineas Chapin, who gave the family name to the surrounding community.

Increased population, peace, and prosperity coupled with growing per capita use of iron led entrepreneurs in the northwest into a burst of bloomery forge construction during the 1740s. Then, hard times arising from the wars with the French colonies and their Indian allies virtually stopped new forge construction in the next decade (see table A1.2).

Since few in the northwest had enough capital of their own to build a forge, entrepreneurs sold shares to partners. Eight of the first forty-five men in Kent (in 1739) invested in ironworks. According to historian Charles Grant, the list of iron men looked like a social register of early Kent.[5] Family and neighbors often took these shares and then traded them among themselves in frequent, complex transactions as their individual fortunes waxed and waned. Additionally, family members often passed forge shares on to their children. Timothy Hosford, partial owner of ironworks at the Great Falls of the Housatonic by 1743, sold a partial interest in 1747 to Charles Burrall (the bridge proprietor; see chap. 2). Asa Douglas joined the partnership in 1749; shortly thereafter the forge was known as the Asa Douglas Iron Works or the South Iron Works. In 1765 Douglas sold half interest in the works and all his land near the falls to Charles Burrall; Burrall acquired the whole interest (it was then known as Burrall's Forge) and in 1772 sold a portion to his son William and another to Russell Hunt in 1773. In 1801 Joseph Coe bought all the forge shares.[6]

Some investors from outside the northwest still participated in bloomery enterprises in the 1740s. Thus, after Ebenezer Barnum and his three sons bought land in East Kent on the West Aspetuck River for a bloomery in 1744, they sold a quarter of it the next year to Thomas Fitch and Robert Walker Jr., already proprietors of the Kent ore bed, to get cash to start the forge. Fitch and Walker remained partners with the Barnums for the next five years.

Widespread changing ownership of forges involved a large segment of the Kent community in ironmaking. The need to share waterpower resources added complexity to enterprises based on this renewable resource. Participants needed to work out explicit agreements that protected their own interests and those of the larger community. When, in 1747, Fitch and Walker, partners in the Barnum forge venture, bought land from Sylvanus Hatch that controlled release of water from North Spectacle Pond, they also concluded an agreement on the release of water to Hatch's gristmill located below the ironworks (fig. 3.2). Hatch was to:

- Build and maintain a dam in good repair so as to form a reservoir to save water.
- Keep his mill in good order so as to use water prudently.
- Draw water only during five months of the year, at his election, except that if he drew water in a particular month, that month must be one of the five.
- Not draw water unless "the Ironworks lay still for six days together."

Fitch and Walker, for their part, agreed not to build a gristmill above Hatch's so long as the ironworks continued and Hatch kept his mill in good repair to serve all customers.[7]

North Spectacle Pond gathered water from surrounding wetlands. However, because of its high elevation and its position at the head of the West Aspetuck River, its watershed was small, making the quantity of

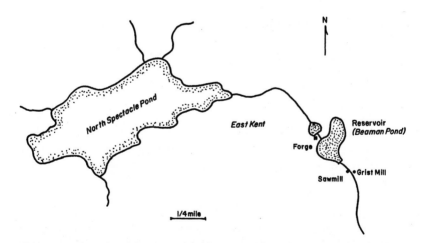

Fig. 3.2. Map showing the locations of works using water from North Spectacle Pond in East Kent. When Fitch and Walker bought land near the outlet of the pond to gain control of a reserve of water for the forge in which they were partners, they gained control of the water supply to the saw and grist mills downstream. They worked out an agreement that required the owner of the gristmill to construct a storage reservoir, now Beaman Pond, to assure prudent use of the available water.

water it could supply during the dry part of the year insufficient for steady operation of the ironworks. The forty-six-foot fall from the pond to the ironworks could have provided the necessary power with only a small flow. However, no one had the capital needed to build a power system that could utilize such a high head until at least the mid–nineteenth century. With the waterwheels they could afford to construct, entrepreneurs like the Barnums had to use the available flow prudently. The agreement with Hatch recognized the need for prudent use and served as a vehicle for avoiding disputes. The agreement illustrates the practical understanding of hydrology among Salisbury artisans.

Non-resident investors generally dropped out of the northwest's iron business within a decade or so, with full control passing to local families. The early colonial investors—Williams, Eliot, and Walker—had started selling out their interests in ironworks, beginning in 1743 with Kellogg's Bantam bloomery forge. Eventually Daniel Rowe of Suffield, an experienced bloomer, bought out the remaining partners in the Bantam forge and made iron until 1792. In 1751 Fitch and Walker sold their interest in the Barnums' East Kent forge for five tons of bar iron. Jonathan Morgan of Kent first bought a share in this forge in 1754; his son Caleb began buying shares in 1765 and gained control in 1801. Various other Kent residents owned shares from time to time. The Morgan forge, as it came to be known, was one of the region's longest lived bloomeries, with Morgan's sons continuing to make iron there into the 1850s.[8]

Some individual artisans combined owning a bloomery forge with shaping the metal they made into finished products. Edward Cogswell, using ore from the Kent Mine, operated a bloomery from 1745 until about 1800 in New Preston, making iron for blacksmiths and manufacturing tools and farm equipment (see fig. 2.5).[9] Bloomeries could be located in the very center of a village. Cogswell's was on the east side of the East Aspetuck River on the north side of the crossroads that marked the center of New Preston.

New settlers in the Salisbury district who speculated in real estate while investing in ironworks could easily get themselves into burdensome debts. Benajah Williams, a resident and initial church member in Salisbury, purchased substantial acreage from Lamb in 1743. Josiah Stoddard came to Salisbury in 1743 from Wethersfield, bought a farm in Lakeville, and made himself prominent in town affairs. He was a captain of the militia and an investor in Maine and New Hampshire lands. William Spencer, who came to Salisbury in 1748 and was a partner with Lamb in real estate deals acquired a share in Ore Hill in 1749. Among Lamb's valuable real estate was the water privilege at the outlet of Furnace Pond. Always ready to sell for a quick profit, Lamb sold this privilege in 1748 to Williams, Stoddard, and Spencer, who used it for their Lakeville bloomery forge.[10] They sold shares to others local residents.[11] Williams, Stoddard, and Spencer soon found that they could not make enough money from the forge to service their debts. John Pell, the surveyor and Ore Hill proprietor, began buying them out with borrowed money, and in 1753 he made himself owner and manager. He, in turn, failed to meet the condition of his mortgage that he deliver 11.5 tons of bar iron per year to his creditors, and Leonard Owen gained possession. Owen subsequently sold the forge to Ethan Allen and partners, who built their blast furnace on or near the site (see chap. 4).[12]

A later generation of bloomery entrepreneurs often managed their affairs better. Angus Nicholson, a Scot who settled in New Milford, purchased the site of a sawmill on the West Aspetuck River in 1768 for a bloomery forge. By 1787 he was able to buy half of the Lanesville forge, the area's first bloomery. Nicholson's property, as listed in a 1792 mortgage, included twenty-three acres of land, the ironworks, dwelling, sawmill, potash house, oil mill, blacksmith shop, nailery, coal house, stables, shops, and miscellaneous buildings.[13]

Bloomery proprietors who eventually converted their forges into manufacturing establishments had the largest impact on the Salisbury iron industry. Richard Seymour of Hartford, a smith and owner of a one-eighth share of Ore Hill, bought property on the Blackberry River in Canaan for a bloomery forge at the 1738 proprietors' auction. Seymour had mastered iron-smelting skills well enough to get his forge into production promptly, and he encouraged John Forbes of Hartford, who had moved to Simsbury to learn smithing, to set up a shop on the north bank of the river across from the bloomery forge in 1746. Here Forbes used Seymour's iron to make tools, chain, wagon parts, nails, and other products that local people needed. Forbes was able to buy Seymour's bloomery in 1751 and take his sons Samuel and Elisha into the business.

Conclusion of the colonial wars with France saw increased demand for iron and for forged products that an ordinary smith could not easily make. Millwrights needed parts for new gristmills, sawmills, and fulling, paper, and oil mills, as well as repair parts for older ones. Shipbuilders wanted larger anchors as they ventured into making bigger vessels. Samuel and Elisha Forbes enlarged their skills and their works to fill all these needs through the 1760s, thereby starting what would prove to be one of the district's most important industries (see chap. 5).[14]

Another increase in demand for iron, and a wave for bloomery building, followed the return of men from military service and more stable economic conditions after 1783. Slitting mills in East Canaan, Litchfield, and Washington (see chap. 5) added to the demand for bloomery iron as they supplied rod to the region's many nailers, including those chained to their anvils in Newgate, the state's prison in East Granby.[15] Once built, these rolling and slitting mills could process iron at a faster rate than the district's existing bloomery forges could supply metal. In the two decades before 1810, district entrepreneurs opened at least thirty forges (see table A1.3). Most prospered, including three forges and a slitting mill that formed a new industrial district in Woodville on the Shepaug River. Here Caleb Hitchcock and Abijah Pratt built a bloomery near the intersection of the boundaries of Litchfield, Warren, and Washington on the west bank of the river in 1783 to smelt ore from South Kent (fig. 3.3). When Forbes & Adam, proprietors of a slitting mill in Canaan, converted the gristmill

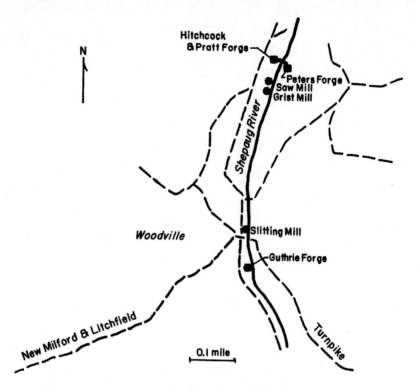

N

Hitchcock
& Pratt Forge

Peters Forge
Saw Mill
Grist Mill

Shepaug River

Woodville

Slitting Mill

Guthrie Forge

New Milford & Litchfield

Turnpike

0.1 mile

Fig. 3.3. The Shepaug River drops seventy-three feet in its 1.2-mile course through Woodville. The large drainage area upstream assured ample flow. Three bloomery forges and a slitting mill shared the waterpower of the Shepaug in Woodville with a gristmill and a sawmill. Since the river provided abundant water, large storage ponds and cooperative use agreements were less needed here than at the works using water from North Spectacle Pond (see fig. 3.2).

downstream of the Hitchcock and Pratt forge to a slitting mill in 1794, they increased the demand for iron. Joseph Guthrie and Elijah Mallory immediately built a bloomery forge a quarter mile farther downriver. Then, in 1799 Eber Peters Sr. of Litchfield added a third bloomery to complete this cluster of works. All three bloomery forges continued making iron until about 1850.[16]

The Woodville forges got their ore at the Kent mine. Usually one of the proprietors would go to Kent to pick up a load of washed ore that the mine manager had graded. He paid for the ore several weeks later by delivering a portion of the iron his forge has smelted. They got in the habit of relying on the mine manager to grade the ore carefully. Careless work by the manager in the spring of 1820 led to complaints by the forge proprietors. When the forge owners forced Samuel Johnson and John Adam Jr., the mine proprietors, to pay refunds for bad ore, Johnson and Adam decided to shift responsibility for quality assurance. They warned the forge owners "that, as the qualities and value of the different kinds of Ore can not be exactly judged until the iron is worked in the Forge, the Purchaser must take the Ore entirely at their own risque, and their own judgement."[17] These difficulties foreshadowed problems that would become severe when the Salisbury district makers concentrated on high-quality iron made for demanding customers, such as the national armories (see chap. 5).

The independent artisans running the Salisbury district's bloomery forges and slitting mills carried on their business within the New England exchange economy that centered on store owners and complex trading networks that functioned with only occasional cash settlements.[18] An artisan might work for the proprietors of a forge or mill and simultaneously on his own account, as Simeon Palmer did from 1797 to 1802. While Palmer operated Wadsworth & Kirby's slitting mill in Bantam, he bought iron on his own account from a Litchfield storekeeper, rolled and slit it, and delivered nail rod to customers who bought from the store. Palmer, Wadsworth, & Kirby, and the mercantile partnership of Elijah Wadsworth & Company all traded nail rods among themselves and the store customers. Difficulties arose in sorting out these complex trades when Palmer went bankrupt in 1802.[19]

As long as demand was strong, the products unsophisticated, and resources abundant, bloomery forge proprietors did not need to worry much about the efficiency of their smelting operations or the quality of the metal they made. After about 1825, as more people entered the trade and customers demanded a more uniform product, bloom smelters had to pay closer attention to their technique. The experience with ore from Kent showed how demand for better uniformity in their product could overtax artisans' craft skills. In the 1830s, bloom smelters in New Jersey and the Adirondack region of New York improved their technique by developing the American bloomery process, which used hot blast to save fuel and a hearth design that helped the bloomery minimize loss of iron in the slag.[20] Archaeological evidence from Salisbury district bloomery sites suggests that Connecticut bloomers had learned to reduce the amount of iron they lost in slag but that they ignored (or never learned about) the new American bloomery process.[21] Bloomery forge operators throve in the Adirondack region by making high-quality specialty iron from the high-grade magnetite ores their region held. In Salisbury, which lacked these ores, the indirect (blast furnace—finery) process offered a better way of making superior quality metal.

The remaining bloomery forges in the district closed in the 1850s, finally displaced by the newer finery forges and puddling works. However, opportunities for independent ironmaking artisans remained open in mining and coaling, since demand for fuel and ore continued strong.

Merchant Capitalists

Bloomery forges could not produce cast items such as hollowware, potash kettles, or the hammerheads and anvil bases artisans needed for their forges. With a blast furnace, an ironmaker could both increase the scale of iron smelting and provide customers with iron castings. However, in colonial times a furnace required an investment of over £3,000, while a fully equipped bloomery forge cost only about £500.[1] Additionally, blast furnace proprietors had to assemble a workforce at least ten times larger than they would need for a bloomery, provide a continuous supply of ore and fuel over a period of months, and create a stable organization to manage the whole enterprise. Bankruptcies among the initial furnace owners in Salisbury showed how difficult these tasks could be.

Philip Livingston, proprietor of the manor across the border in New York, had access to capital that no one in Connecticut could match. In 1743 he had a blast furnace built at Ancram, in the western part of his manor, to smelt ore brought over the Connecticut border from Ore Hill.[2] Nineteen years elapsed before Ethan Allen and his partners managed to launch the first such venture in the Salisbury district.

Blast furnace proprietors needed an absolutely reliable water supply, since they had to keep their furnace's running continuously for months. The place closest to Ore Hill that could provide such a supply was the outlet of Wononskopomuc Lake, soon to be renamed Furnace Pond. Water from this lake could run the approximately two-horsepower air pump needed for an eighteenth-century blast furnace throughout the year.

A thriving colonial economy in 1760 encouraged John Hazeltine of Massachusetts—founder of the town of Upton, dealer in lands, and proprietor of the bloomery forge in Sutton—to send his son Paul to scout out investment opportunities in mineral-rich and underpopulated Salisbury.[3] Here Paul met Ethan Allen, later to gain fame at Fort Ticonderoga, who had just sold his Cornwall farm for fifty pounds. The entrepreneurial Allen watched high-grade iron ore from Salisbury's mines hauled over to the Livingstons' Ancram furnace in neighboring New York and learned that anyone who wanted a potash kettle or other cast-iron product had to have it made in New York or Massachusetts.

Allen saw opportunity here. Paul Hazeltine had access to capital. Ore, fuel, and a source of waterpower were at hand. If he could get Hazeltine to invest, all Allen needed was someone who knew how to build and run a furnace. The brothers Samuel and Elisha Forbes, masters of a forge across the Housatonic River in Canaan could do this, and they had some profits ready to put into a new enterprise.

Allen, John Hazeltine, and the Forbes brothers agreed on 16 January 1762 to build "A Furnice to Run Iron Pig mettle" and prepare enough ore and fuel for two months of smelting by 5 September. John Hazeltine put up half the capital, the Forbes three-eighths, and Allen the remaining eighth. John came to Salisbury to manage the enterprise.[4] The partners bought Leonard Owen's bloomery forge at the outlet of the lake (fig. 4.1), along with his ore rights and 365 acres of woodland on nearby Mount Riga for coaling (with chestnut excluded because of its use for split rails) for £430. Samuel Forbes built and blew in the furnace, following the design used a hundred years earlier at Saugus and used again at Mount Riga (fig. 4.2), and soon had it up and running.[5] The partners hired a pair of

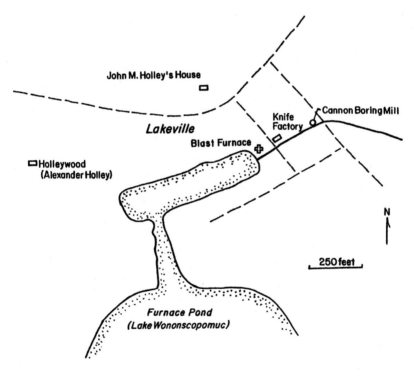

Fig. 4.1. Location of historic sites in the village of Lakeville in the town of Salisbury. Allen and partners built their blast furnace at the outlet of Lake Wononscopomuc (later Furnace Pond). The cannon-boring mill built during the Revolutionary War was probably located downstream of the furnace, as shown. John M. Holley's house still stands on the hillside above the furnace. Alexander H. Holley built his first knife shop on the furnace site and later replaced it with a brick factory (standing today) across the street. His home, Holleywood, overlooks Furnace Pond.

Fig. 4.2. The blast furnace built on Mount Riga in 1810. The furnace built at Lakeville was probably similar in design. (Photo by William Sacco.)

keepers to watch the furnace day and night, three "pounders" to break up the ore and flux, two fillers to charge the furnace, a banksman to unload ore, two coal stockers, and a jobber to clear the casting floor. A clerk kept the books, while a carpenter and blacksmith made necessary repairs.[6] With the ore diggers, colliers, and teamsters, upward of sixty hands worked in the enterprise, making up the community that people began to call Furnace Village.[7]

Allen's managerial talent (and his sense of civic responsibility) fell far short of his entrepreneurial skills. In three tumultuous years in Salisbury he repeatedly demonstrated his physical prowess by assaulting his neighbors, shocked and (privately) entertained the townspeople with his rich command of profanity, and soon parted company from his partners. John Hazeltine and Samuel Forbes left the partnership by selling their furnace shares to Charles Caldwell of Hartford and his brother George of Salisbury. The Caldwell family owned property in Hartford and several other towns and ran a store in Salisbury. Allen sold his interest to George Caldwell in October 1765 and left town, making the Caldwells 25/32 owners.[8]

Low prices for pig iron after the French and Indian Wars, competition from Livingston's Ancram furnace (which could ship iron to the Hudson at lower cost), and inept management led to losses as the Caldwells attempted to run the furnace enterprise. In 1767 they appealed to the General Assembly for a loan of £1,200 to cover their operating costs. When this proved insufficient, they mortgaged part of the property and again appealed to the assembly. Meanwhile, they had gotten in debt to Richard Smith, a Boston merchant and shipowner knowledgeable in the iron trade, for store merchandise. To settle this debt, the Caldwell brothers sold Smith a 17/32 furnace share in 1768. Further difficulties followed, and Smith gained control of the Caldwells' furnace property with a payment of £832. The Caldwells' ironmaking venture wiped out the family wealth and left George jailed for the furnace debts.[9]

It took the managerial skills and financial resources of merchant-capitalist Richard Smith to make a success of the furnace enterprise. Smith lived in Boston and had his agent, Jared Lane, conduct day-to-day business in Salisbury. Lane arranged boarding for the furnace workers. He contracted with individual Salisbury artisans to make charcoal, dig ore, haul the ore and fuel to the furnace, and haul the pig iron and the kettles the furnace artisans cast to individual customers or to Hartford for shipment to Boston for export. Representing the interests of a non-resident capitalist in a community that had seen a succession of failures in the iron trade was a task that led Lane into altercations with local property owners and artisans. By 1773 he was fed up, writing, "I think there appears the least generosity and honesty in the people about this place of any people that ever I see."[10]

Once he had a blast furnace, Richard Smith could make wrought iron by the indirect process instead of by bloom smelting. He could then place a forge far from the mines, at a superior water privilege and close to abundant fuel, where the cost of hauling ore would make a bloomery uneconomic. Smith built his finery forge in the Robertsville section of Colebrook, distant from other ironworks that might compete for fuel, at the head of the gorge on the Still River, one of the best water privileges in the northwest.[11] The Still received water from Long Pond (now Highland Lake) in Winchester (fig. 4.3).[12] In 1771 Smith made a deal with the proprietors of the town of Winchester that allowed him to draw down Long Pond as much as a foot and a half to supply water to his forge, and he had a log dam built at the outlet of the lake to control the release of water. He had to make arrangements with other waterpower users along the stream, as had the proprietors at the Fuller Forge in Kent (see chap. 3).[13]

Since no one in Connecticut knew the fining technique, Smith turned to New Jersey, where artisans at the Union Iron Works had been fining iron since the 1750s.[14] Here he recruited Jacob Ogden to run his new forge. With the fining process, Ogden could make the highest quality wrought iron, which was difficult to produce in a bloomery. Smith built his business by touting the quality of his products. For example, a 1771 advertisement in the *Boston Evening Post* for a potash kettles "cast in Salisbury from best mountain ore" claimed them to be superior to those made from bog ore, which were liable to crack when first used. Additionally, owners could sell worn out kettles for a high price for recycling at a finery forge because of the established reputation of Salisbury iron as a starting material for fining.[15]

Smith succeeded through his knowledge of the iron trade and his

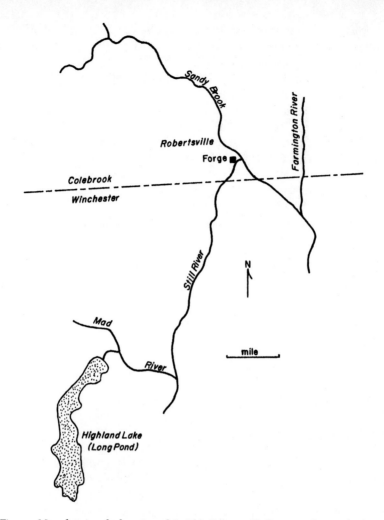

Fig. 4.3. *Map showing the location of Smith's Robertsville forge. By placing the forge at the head of the gorge where the Still River drops forty feet to join Sandy Brook, and arranging to draw water from Highland Lake (then Long Pond), Smith assured himself of a superior fall and flow of water to drive the forge hammers.*

ability to hire skilled artisan-managers for both the Lakeville furnace and the Robertsville finery forge. However, his position as a Boston merchant engaged in trade with British agents made him vulnerable to Patriot suspicion of Tory sympathy. He wrote that he met with "no Insults in New Hampshire" in a February 1775 journey to Portsmouth, where he dispatched cargoes of iron from the forge, and thought that his refusal to pay duty on a cargo to tea proved to Bostonians that he "was no Enemy of America." However, he was anxious to collect all debts as "these are no times to trust property."[16] By July 1775 he had decided to wait out the revolutionary unpleasantness in England and Jacob Ogden ran the forge in Robertsville on his own account through the war years. Governor Trumbull sent a committee to inspect the Lakeville blast furnace in January 1776; on its recommendation, the Council of Safety seized the works for cannon making.

Since no one in Salisbury had ever cast cannon, the committee recruited Lemuel Bryant of Middleborough, Massachusetts, an experienced cannon founder, along with molders David Carver, Zebulon White, and David Oldman to cast cannon at the Lakeville furnace in 1776. Salisbury artisans contracted to supply charcoal to ironworks.[17] Local residents Medad Parker and Joel Camp worked as pattern makers; Daniel Forbes was chief banksman; Edward Whitcomb acted as foreman filler. Bryant's men cast the cannon solid and then bored them out in a mill built for the purpose near the furnace and powered by water in Furnace Brook (see fig. 4.2). By the end of 1776 they had cast 156 tons of cannon in sizes from three-to eighteen-pounders, 38 tons of shot, and about 20 tons of other castings.[18] By 1779 the Council of Safety found the furnace too expensive to operate and put it out on leases to a number of different individuals.

At the war's end, Connecticut ironmasters faced disrupted markets and the problem of recruiting the stable force of artisans needed to operate a blast furnace. In 1783 the state returned the Lakeville furnace to Smith, who promptly sold it to Elisha Buell. Although Buell and his successors, Joseph Whiting and William Nelson, had difficulty raising the capital they needed to put the furnace into good condition after lack of maintenance during the war, they got it back into production.[19] In 1798 artisans at the Lakeville furnace cast cannon ranging in size from four-to thirty-two-pounders for the New York militia and the navy. Some went to the Battery in New York City, and the rest to Commodore Truxton for use at sea.[20] However, Whiting and Nelson made their primary business supplying pig iron to finery forges and the cast-iron items that district artisans needed.

Richard Smith had prospered by making the Lakeville furnace part of an established mercantile business and by vertical integration. The Holley family succeeded by adopting the same plan. Luther Holley, who moved to Salisbury from neighboring Sharon, had made enough money trading Connecticut merchandise to New York City by way of the Hudson River to add ironmaking to his mercantile business. In October 1799 he purchased 1,800 acres of land; the blast furnace, water privilege, store, and gristmill at Lakeville; a half interest in an ironworks and a water privilege on Mount Riga; a tenth interest in Ore Hill; seven dwellings and barns; and two cannon-boring mills—all from William Neilson. Luther sold part of the property and took his son John M. Holley into partnership as L. Holley & Son for trading and ironmaking. The partners prospered, allowing Luther to retire from an established business in 1810.[21]

Upon Luther's retirement, John Holley formed a partnership with John C. Coffing—a Southbury native who had established himself as a capable businessman in Salisbury—to run the ironmaking component of L. Holley & Son's business. The partners' account books show how they ran the enterprise in their first year.[22] While neither Holley nor Coffing had any experience in ironmaking, they could hire experienced artisans in Salisbury. On 23 April 1808, Ebenezer Brinton, helped by David Blodgett, was tearing out the damaged furnace hearth. A week later Ebenezer, James Brinton, Samuel Smith, Samuel Davis, and several Salisbury farmers took their wagons to Poughkeepsie to bring back nine tons of firestone (the special quartzite used to rebuild the hearth) that had been shipped up the Hudson River for them.

On 5 May Luther Holley & Son started sending a quart of rum every few days from their store to the men at the furnace. By then Ebenezer (at $1.00 per day) and Joseph Brinton (at $0.83 per day) were rebuilding the hearth. They fitted blocks of stone weighing up to 150 pounds tightly together to form a container that would hold a ton of molten iron and slag. They finished on 28 May, repaired the flume that brought water to the bellows wheel, and had the furnace ready to go. Coffing & Holley started buying ore, and teamsters brought in loads of charcoal. Other local artisans supplied flux, clay, and molding sand.

On 3 June Coffing & Holley hired Ebenezer Brinton as founder (in charge of the furnace at $50 per month); Micah French as banksman (in charge of the stocks of ore, fuel, and flux at $11 per month); William Hopkins and Elijah Easty as guttermen (working the pig bed, $11 per month); J. Everest, John Montgomery, and J. Brinton as fillers (charging the furnace, $15 per month); six laborers as ore pounders; and Joseph Brinton and two others for general labor about the furnace. They made up about the same size crew as had worked the furnace forty years earlier. To prepare the ore for the furnace, workers broke it up and placed it in layers alternately with charcoal to make a heap three to four feet high, which they then burned for several hours.

It took Brinton about a week to get the furnace hot enough to begin charging ore. Once it was going well, he was able to make about two tons of iron a day (the same rate at which Lane shipped pig from the furnace in 1771).[23] The furnacemen either ran the iron into a bed of sand to make pigs or dipped it from the furnace hearth with hand ladles to pour into molds for kettles or other cast products. The pig they made sold for $35 per ton, open mold castings for $70, and flask castings for $80. Brinton and his men kept the furnace going for 196 days, making a total of 440 tons of iron from 1,000 tons of ore and 800 tons of charcoal. L. Holley & Son continued to send over "rum for the furnace" every few days until early December, when they substituted gin and doubled the quantity. Work continued as usual on Christmas Day and New Year's Day.

Salisbury now had many of its men working at its mines, forges, and furnaces. They were making a visible impact on the environment. Forty-seven different individuals sold ore to Coffing & Holley during the 1808 furnace campaign. Each blasted loose and dug out ore at his own pit at Ore Hill, alone or with a few helpers. They sold ore for $2.79 per ton, and paid a $1.04 per ton royalty to the mine proprietors. Teamsters hauled one-ton wagonloads the two miles from Ore Hill to Furnace Village every day. During the 196 days of the blast, a typical miner would dig out about

500 cubic feet of ore and overburden, leaving the hill pockmarked with individual miners' pits, each with its pile of spoil. One observer thought it looked like giant woodchucks had burrowed all over the hill.

Ore digging offered anyone in Salisbury part-time or off-season work and the opportunity to accumulate enough capital to begin a business. Josiah Read, a farmer near Ore Hill known for his hard work and enterprise, used his mine earnings to become a partner in the Wassaic blast furnace just over the border in New York. Lee Canfield, one of the ore diggers, eventually made himself a forge owner and bank president.[24] Seneca Pettee was building his own blast furnace on Mount Riga while digging ore for cash. In 1808 Salisbury had merchant capitalists with local roots, a fluid economy, skilled artisans taking on varied tasks, and opportunities for upward mobility.

Luther Holley & Son sold 376 of the 400 tons of iron the furnace made in the 1808 campaign to the Rockwell brothers and to James Boyd, finers in Winsted. The Rockwells, Boyd, and others were then so perfecting their technique for making iron of the highest quality that they would soon be the preferred suppliers of iron to the national and private armories. Nevertheless, the twenty-four tons of castings made at the Coffing & Holley furnace in 1808 were an essential component of the local industrial economy. They included screw boxes, gudgeons (bearings) for Forbes & Adam, potash kettles, and twelve forge plates and an anvil of 1,330 pounds that James Boyd needed to enlarge his forge. Luther Holley & Son also supplied the community with daily necessities: oats, rye, wheat, tobacco, rum, brooms, tallow, axes, shovels, powder, shot, shoes, cloth, sugar, tea, and, for the miners, blasting powder. In 1815 Holley and Coffing included construction of a general store on the furnace premises in their partnership agreement. Most Salisbury proprietors included a mercantile component in their ironmaking enterprises through the first half of the nineteenth century.[25]

Ironmakers considered a blast furnace a fickle mistress, and with good reason. Variations in the composition of the ore; the quality of the charcoal; the moisture content of the air blown into the hearth; the placement of ore, fuel, and flux by the fillers; and the condition of the furnace lining all affected the quality of the iron made. William Dunbar of Boston complained bitterly in 1810 about eight tons of pig he ordered from Holley & Coffing. The partners explained that stormy weather and bad roads made it difficult to get the iron to Hudson, the port on the river where a ship would pick it up for Boston. When Dunbar complained that much of the iron that did arrive was too soft, the partners replied, "It is not always possible to make a furnace go as you want to. . . . the most zealous exertions sometimes fail to produce the results exactly that you want."[26]

Despite problems such as those with Dunbar, the partners built a reputation such that Commodore William Bainbridge wrote to ask them to cast cannon and shot for the navy in 1813. J. M. Holley replied that Salisbury had the best ore in the country available in unlimited quantity, that the partners were making 500 to 700 tons of iron a year, and that they were about to deliver 200 tons of conduit pipe to Albany. However, they would need to build another furnace adjacent to one of the two they operated since neither furnace could hold enough metal to cast a large cannon. They would want a large contract and an advance of cash to undertake this. Holley's assertion that, while he and Coffing had never cast a cannon, they could easily hire molders with the requisite skills, illustrates their reliance on artisans to solve problems of technique. The next year Decius Wadsworth, chief of the army's Ordnance Department, again asked Holley & Coffing to undertake cannon founding. The partners now had more than enough business in hand, and declined.[27] They wanted to concentrate on smelting pig iron for the district's finery forges, which had found a product that ironmakers elsewhere could not match: gun iron, the material used at the national and private armories for musket and rifle barrels (see chap. 5). The district's forges supplied the armories with gun iron until the mid-1850s.[28]

Holley & Coffing, along with the satellite partnership Holley, Coffing & Pettee formed in 1810, established themselves as Salisbury's principal firm in the iron trade, a position they would hold for the next quarter century. The resources Luther Holley and his son accumulated through their mercantile trade and an expanding economy helped Holley & Coffing succeed where others had failed.

Salisbury ironmasters increased their demand for pig iron beyond the capacity of the Lakeville furnace as they built new finery forges in the early decades of the nineteenth century (see table A1.3). To meet this demand, Seth King of Salisbury decided to build a blast furnace on Mount Riga. The ponds near the top of the mountain offered the reserve of water that could assure uninterrupted furnace operation. However, teamsters would have to haul ore nearly a thousand feet up the mountain from the Ore Hill and Davis mines below. With John Kelsey as partner, King bought land and water rights for a furnace in 1802 and 1803. They used a site near the outlet of South Pond, where Abner and Peter Woodin had set up a bloomery about 1781 that had passed through several owners to Luther Holley in 1799 (fig. 4.4).

King and Kelsey's tribulations show how difficult it then was for individuals lacking an established business to raise the capital needed for a blast furnace. The partners made complex arrangements with other Salisbury entrepreneurs to raise money. With financial backing from Peter Farnham, Amos Moore, and David Waterman, they replaced the old timber dam at South Pond with a stone one and commenced construction of their furnace. By December 1803 they were unable to pay the debts they incurred buying land and had to release their holdings, including the unfinished blast furnace, to their creditors, Waterman, Farnham, and Moore. Waterman then gained control of the entire property. In 1802 Luther Holley sold his half interest in the ironworks to Elisha Sterling, an attorney in Salisbury; in 1804 Waterman sold his half interest in the forge and furnace to Sterling, who then held the entire property. Sterling immediately sold the dam, forge, and furnace to two experienced ironworkers from Norfolk County, Massachusetts, Joseph Pettee of Foxborough and

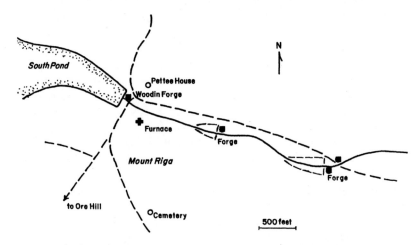

Fig. 4.4. Map of the ironworks on Mount Riga. Woodin's forge and the blast furnace drew water from South Pond, which in turn was fed by North Pond (not shown). Holley & Coffing's upper and lower forges each had a dam to form a pond holding a reserve of water. The furnace and forge proprietors chose this remote location high above the iron mines because of its water and forest resources.

Willaby Dexter of Walpole, who moved to Mount Riga in 1805.[29] Dexter worked as founder at the Lakeville furnace for L. Holley & Son from 1805 to 1807; Pettee dug and sold ore in 1808.[30] Pettee and Dexter finished the furnace and sold it to Holley & Coffing in 1808. They operated it with Joseph Pettee as a partner, and with Joseph and his brother Seneca managing through 1838, when Joseph died. Joseph's son, William J. Pettee, remained as manager until 1845. New owners abandoned the furnace in 1856.[31]

Leman Bradley built Salisbury's third blast furnace in 1812, he using a site at the base of the Great Falls on the west side of the Housatonic River (see fig. A1.1). He immediately rebuilt it after a fire in 1814.[32] Benjamin Silliman, who visited the furnace during his tour of the northwest in 1817, described it as about twenty-five to thirty feet high and ten to twelve feet in diameter; it was four to five feet in diameter at the top. Water-powered bellows supplied the blast. The mine at Ore Hill supplied most of the ore, and adjacent New York mines provided the rest.[33] W. C. Sterling and Lee Canfield formed Canfield, Sterling & Company to run this furnace sometime before 1837.[34] The partnership of Canfield & Robbins finally abandoned it in 1857.

As American manufacturers recovered sales lost to the British imports that flooded the country at the end of the War of 1812, they increased their purchases of bar iron. The large market for Salisbury finery iron in turn increased demand for forge pig. The immense piles of slag and refuse around the Lakeville furnace noticed by Benjamin Silliman on his 1817 visit indicate substantial production.[35] The Mount Riga furnace had been built explicitly to supply forge pig, and the Bradley furnace added to the supply.

Salisbury entrepreneurs responded to demand from the forges and the growing New England industrial economy by launching a new round of blast furnace building starting in 1823. By 1826 they had added five to the three they already had: Macedonia in 1823, Sharon Valley in 1825, Kent in 1825, Bulls Bridge in 1826, and Chapinsville in 1826. W. C. and F. A. Sterling and Abiel Chapin financed the furnace erected at the site of Camp's forge. About 1830 Holley & Coffing built the Lime Rock furnace to operate with their new finery forge. Other entrepreneurs added four more furnaces before the panic of 1837 arrested further construction (see fig. A1.1). The relatively small, cold-blast furnace design they used facilitated smelting pig with the low carbon and silicon content wanted by finery forge proprietors. The district blast furnace builders followed designs well tried by experience.[36]

Completion of six furnaces in this short time (where only three had been built in the previous sixty years; see table A1.2) shows that many Salisbury artisans now had the skills necessary for furnace construction and operation. The district's merchant capitalists relied on these artisans to handle the technique of ironmaking. However, relatively few people in the district were specifically either artisans or capitalists. Artisans who managed their own businesses, or capitalists who worked with their hands alongside their men made up the bulk of Salisbury's ironmakers.

Artisan-Entrepreneurs

The adventurers and colonial investors who initiated ironmaking in the Salisbury district hired artisans to run bloomery forges and make products such as nails and hardware needed by settlers in the new lands. Within a few years artisan-proprietors began making these products in their own forges (see chap. 3). Then a new generation of entrepreneurs with both artisanal and managerial skills began making and selling sophisticated products to distant as well as local customers. New opportunities in the iron trade opened for them in the years before the Revolution. As they exploited these, they transformed the region's ironmaking into a key component of the colonial industrial economy.

In 1739 Richard Seymour, a Hartford smith, started ironmaking in East Canaan by building a bloomery forge on the Blackberry River. He smelted iron ore from the recently-opened mine at Ore Hill and forged products needed by the area's settlers. A few years later he took on John Forbes, also a smith from Hartford, as a partner. Forbes's sons, Samuel and Elisha, learned smithing from their father and the art of bloom smelting from Seymour. By 1760 they had transformed the business from one serving a local market to industrial production by expanding sales throughout southern New England and concentrating on specialized products such as sawmill gudgeons and cranks, gristmill spindles and rinds, clothiers' screws for fulling mills, spindles for paper mills, screws for paper presses, gears, ship's anchors weighing up to a thousand pounds, bellows pipes, logging chains, gun barrels, forge trip hammers, and nail rod.[1]

To meet the growing demand for their products, the Forbes brothers built a second bloomery forge in Canaan in 1759 and another in Norfolk in 1760; in 1761 they purchased a share in the Chatfield ore bed near Ore Hill, first opened in 1740. In 1760 they were selling forgings and mill machinery to customers throughout southern New England. They joined Allen and Hazeltine in building the region's first blast furnace (see chap. 4), opened a general store in East Canaan, and built grist and cider mills.

Samuel Forbes advanced the forge business by guaranteeing his anchors (he would replace one that broke within a year of being sold) and hiring commission men to sell his products in major East Coast cities.[2] He standardized specifications for mill machinery to facilitate sales to distant customers. From 1773 through 1777 Forbes made and repaired helve hammers for Robert Livingston Jr.'s forges in New York. He found it difficult to get good steel for the hammer facings, a problem that would continue to plague manufacturers into the next century.[3]

In 1779 John Adam Jr., son of a Massachusetts forgemaster, visited Samuel Forbes to buy rolls for his father's slitting mill. John got his rolls and also found that Samuel had an attractive daughter, Abigail. The next year John and Abigail married. At the same time, John formed a partnership with Samuel to run the Canaan forge, which by then employed fifty hands (fig. 5.1). Adam brought technical knowledge of iron metallurgy and rolling mills along with sales ability to the Canaan enterprise. Forbes & Adam began making parts for mill machinery to numerical dimensions specified in customers' drawings, anticipating one of the techniques that mechanics would later use to make Connecticut the cradle of the American system of manufactures.

The rolling and slitting mill John Winthrop built at his 1642 ironworks closed with the abandonment of the Saugus project, leaving New England without a source of machine-cut nail rod. Decades elapsed before Americans solved the problems of building a reliable slitting mill on their own. The Union Iron Works in New Jersey had one making about a hundred tons of nail rod per year sometime before 1770.[4] Samuel Forbes had started work on a mill in 1778. Aided by Adam's expertise, the partners perfected their technique in the mill they completed in 1785. With it they could add substantial value to bar iron from their forge: the mill could convert $100 worth of bar iron into nail rod worth $133 or rolled iron that sold for $150.[5]

Ironmaking entrepreneurs and artisans in Connecticut exchanged technical information freely. Daniel Rowe and Roger Cogswell of Litchfield visited the Forbes & Adam rolling and slitting mill to learn how to build one of their own, which they had operating in Bantam in 1789.[6] Five years

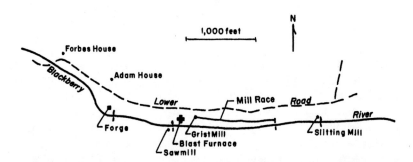

Fig. 5.1. *Forbes & Adam used four dams along the Blackberry River to provide power to each component of their industrial complex. A race north of the river brought water to their gristmill and to the blast furnace built later by S. F. Adam. The other works were located at their respective dams. (Based on a sketch map by Samuel Forbes, courtesy of Dr. William Adam.)*

later the demand for nail rod exceeded the capacity of both mills. Forbes & Adam had used all the good water privileges on the Blackberry River (fig. 5.1), and they had to look for a mill site elsewhere. Hitchcock & Pratt's bloomery forge in Woodville made bar iron that could supply a slitting mill. Here the Shepaug River had unused power potential. Forbes & Adam built their second rolling mill a quarter mile below Hitchcock & Pratt's forge on the Shepaug. In 1799 Joseph Guthrie and Eber Peters added to the supply of bar iron with the two new bloomeries they built in Woodville (see chap. 3 and fig. 3.2).

In addition to slitting iron into nail rod, mill proprietors could roll strips or plates thinner than artisans could make with the trip hammers used at the Salisbury forges. When William Bakewell of New Haven invented a power-driven machine to cut nails from iron strips in 1807, Forbes & Adam supplied him with rolled iron he needed from their East Canaan mill.[7] The partners also used their rolling mill to supply other ironmakers with essential equipment: they made a bellows pipe four feet long, nine inches inside diameter at the large end, and six and a half inches diameter at the other, of plate a quarter-inch thick for the new blast furnace David Waterman was starting to build on Mount Riga in 1803.[8]

Forbes and Adam depended on Salisbury's community of skilled artisans working as independent contractors to staff their forge. These men took on varied tasks, as needed: between June and August 1805, William Foxon made gudgeons, steeled a shaft, turned gristmill spindles, forged and turned fulling mill and sawmill cranks, made a printing press pin for John Roberts of Hartford, turned a clothiers' crank and three bearings, and made pins (screws) for a steamboat and screw plates for a comb factory. He and his son made new rolls and did general repairs at the slitting mill. Other artisans at the Forbes & Adam forge made blades for the saws that cut stone slabs at Marbledale, numerous ship's anchors, and, when called for, just about any other iron product that could be forged or rolled.[9] Each of the artisans working for Forbes & Adam, except a few on day wages, negotiated an agreement with the partners.

With the deaths of John Adam Jr. in 1826 and Samuel Forbes in 1828 (at age ninety-eight), management of their forges, gristmill, and farms in Canaan passed to two of John's sons, Samuel Forbes Adam, a graduate in the Yale class of 1803, and Leonard Adam. The inside-contractor artisans continued to work in their established way while Samuel moved what was now the Adam works away for total dependence on the sale of bar-iron products. In 1832 he built a blast furnace adjacent to the family gristmill.[10] As at the forge, he relied on individual artisans to supply most of the services needed at the furnace: individuals working as miners or teamsters (or both) supplied ore from the Davis mine (which Samuel Forbes had acquired in 1796) in wagonloads of twenty to thirty hundredweight each. Most made one delivery a day; some managed two. Each ore digger rented his own pit from the mine proprietors and paid royalty on the ore removed. The colliers, who supplied the furnace fuel, continued as independent contractors.[11]

John Adam III, the second son in the Adam family, was interested in real estate transactions and investments. In 1832 he sold the site of the paper mill Forbes & Adam had built at the Great Falls of the Housatonic River to a group of outsiders led by the Ames family of Massachusetts. Over the next decade Horatio Ames would be the first to bring new ironmaking techniques to Salisbury, ones that surpassed the traditional methods the Adam heirs continued to use at East Canaan. In 1842 William Adam, the family farmer, began selling off his land for the new village of Canaan that developers were building around the Housatonic Railroad's depot.

The bloomery and sawmill that Russell Hunt and Asahel Beebe of Canaan built along the Hollenbeck River the same year John Adam came to Canaan formed the nucleus of another Salisbury forge run by artisan-entrepreneurs. Since Beebe soon withdrew from the partnership; people called the works Hunt's Forge and the community that surrounded it Huntsville, in the town of Canaan (fig. 5.2). Russell's sons joined the business, making small forgings such as hinges and other home hardware from iron they smelted in their bloomery.

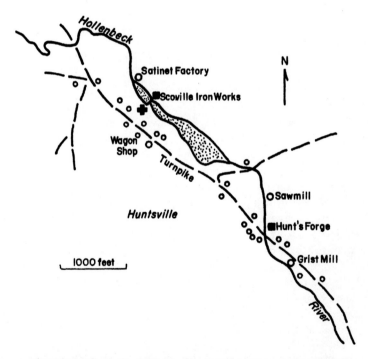

Fig. 5.2. The industrial complex at Huntsville used the eighty-two-foot fall of the Hollenbeck River in a distance of 1¼ miles for a gristmill and a sawmill, Hunt's forge, the Scoville blast furnace, and a textile mill. The Canaan & Litchfield Turnpike offered a good route toward the Hudson River ports, or south to Litchfield, where other turnpikes led on to Hartford.

In the first decade of the new Republic, foreign trade, principally with the West Indies, began to generate wealth that Connecticut merchants could invest in new ventures. Establishment of banks by 1792, and of insurance companies after 1794, facilitated expansion of industry. The Hunt brothers moved into this growing market by making industrial products such as large ship's anchors, screw-press screws, and machinery forgings for Robert Fulton's *Clermont*, launched in 1807. Benjamin Silliman visited the Hunts' forge and screw works in 1817 (fig. 5.3). On arriving at the Hunts' tavern in the evening, he encountered a visitor singing songs of the exploits of the U.S. Navy and discovered that the Hunts were making anchors for the navy's seventy-four-gun ship of the line *Franklin*. The next morning Silliman watched artisans smelting iron into blooms weighing up to 300 pounds. They lifted these from the forge hearth with tongs suspended from a wooden crane and swung them under a 600-pound, water-powered helve hammer for shaping. Forgemen welded the blooms together to make anchors weighing up to twelve tons, after which nine yokes of oxen hauled the anchors to the Hudson River for shipment to the navy yard.[12]

Connecticut entrepreneurs began making goods in factories with division of labor and power-driven machinery starting with metal products in 1798, cotton cloth by 1806, parts for steamboat engines in 1807, and woolens by 1810.[13] The new industries in the nearby Connecticut Valley, the center of burgeoning American precision manufacturing, drew on the Salisbury forges and furnaces for the iron parts for their machinery. The makers of edge tools, firearms, and other metal products created new demand for the high-quality iron. Forbes & Adam, the Hunts, and other Salisbury entrepreneurs made their ironworks an essential component of New England's industrialization. Eli Whitney turned to Forbes & Adam for help with his new armory, and beginning in December 1798 the part-

Fig. 5.3. Drawing of Hunt's forge made by a daughter of the Hunt family. (Courtesy of the Falls Village–Canaan Historical Society.)

ners supplied gun barrels and the machinery Whitney needed to begin mechanized musket making for the federal government.[14] Whitney continued to call on them, particularly when he wanted metal of the best quality: in 1808 he ordered rolled plates to make into bottles for Benjamin Silliman's mineral water establishment. Other customers found the high-quality iron made in Salsibury essential for their manufacturing operations. By 1818 Holley & Coffing was supplying bar iron to J & J Townsend's rolling mill in Albany, to the Columbian Foundry in New York City, and to the Springfield Armory, among others. The armory had its metal sent to Forbes & Adam to be rolled into plates.[15]

Forbes & Adam, Holley & Coffing, the Hunts, and other Salisbury iron-makers met the new manufacturers' demands with the skills of artisans who had mastered the smelting and forming techniques already established in the district. Bloom smelters made high-grade metal most easily from pure ore. They had neither chemical analyses of ore nor knowledge of the specific constituents that would degrade or improve the quality of their iron. Instead, a bloomer had to smelt a sample of ore and then test the resulting metal. When he found ore that gave good results with his technique, he had to depend on the mine to continue yielding comparable ore. We saw in chapter 3 that Salisbury ore was not uniform and that smelters lacked adequate methods of quality control. Operators of forges running on the American bloomery process in the Adirondack region owed much of their success to the uniformly phosphorus-free ore they got from some of that region's mines. Salisbury lacked ore of the unusual quality found in the Adirondacks. However, Salisbury artisans, beginning with the Rockwell family, found that they could use the fining technique introduced by Richard Smith at his Robertsville forge to make the more nearly uniform metal the new manufacturing industries wanted from pig smelted from Salisbury ore.

Samuel Rockwell and five of his sons (Timothy, Solomon, Reuben, Alpha, and Martin) built two finery forges in Colebrook beginning in 1789, as an addition to their sawmill, gristmill, and potashery enterprises (fig. 5.4). The Rockwells were artisan-entrepreneurs in the most remote and last settled town in the northwest, where industry rather than agriculture offered the best hope of gaining wealth. After Samuel and Timothy died in 1794, the family carried on the business as Solomon Rockwell and Brothers. They chose the fining process over bloom smelting for their forge because it was cheaper to haul pig iron to Colebrook than ore, and they subsequently honed their skills to take full advantage of the fining technique to make the best grades of iron at finery forges.[16]

The Rockwells expanded the range of products made in the Salisbury district by building a cementation furnace for steelmaking sometime before 1800. Here they turned to southeastern Massachusetts for technological expertise, as Samuel Forbes had done earlier in recruiting John Adam: a Mr. Jencks of Taunton came to Colebrook to supervise the operation of the steel furnace. The Rockwells found a good market for their blister steel but, like other Americans, lacked the expertise and experience

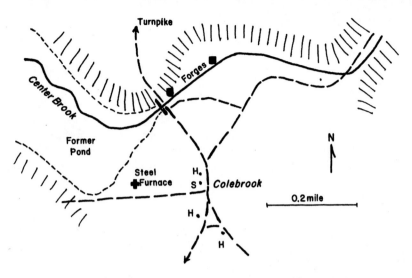

Fig. 5.4. *The Rockwells built their two finery forges along Center Brook, which plunged down from the pond they made with a dam just above the bridge for the Waterbury Turnpike. They placed their steel furnace (which did not need waterpower) west of the village. The Rockwells' houses (H) and store (S) still stand in the village.*

needed to match the high quality achieved by England's Sheffield steel-makers. Sheffield continued to supply nearly all the high-grade steel used by Americans. Artisans at the Collins axe works welded Sheffield steel bits onto axe polls made of Salisbury iron until 1865, when Collins began making its own steel.[17]

Francis W. Rockwell (Reuben's grandson) described from memory the steel furnace at work as he had seen it as a youth. The furnace, made of firebrick and firestone, had an arched top. Daniel Cobb and Phineas Williams (who had replaced Mr. Jencks) packed iron bars covered with charcoal in chests that they capped with clay or sand to exclude air. They put the chests in the furnace, kindled a wood fire in the firebox below the chests, and kept the fire burning evenly by adjusting vents at the ends of the furnace. It took eight to ten days for enough carbon from the charcoal to diffuse into the red-hot iron to complete its conversion to steel. As the conversion neared completion, Cobb and Williams would draw out a bar from time to time to check the progress of carburizing. When the appearance of the fracture at the end of a broken bar indicated that sufficient carbon had diffused into the iron, they allowed the chests to cool for about a week, then removed the bars and took them to one of the Rockwell forges for hammering. The furnace, which cost about $2,000 to build, was housed in a large wooden building.[18] It was still operating in 1846, when it made fifty tons of steel.[19]

James Boyd & Son, the scythe-making entrepreneurs in Winsted, began to compete with the Rockwells in 1832 by erecting a cementation furnace for steelmaking. Most of their steel went into making pitchforks. The Rockwells managed to make steel that was just as good at lower cost, so Boyd & Son abandoned their works after a few years.[20]

The Rockwells encountered an unexpected environmental problem with the waterpower system that ran their Colebrook forges. They had made a forge pond that flooded adjoining lowlands by placing a dam across the stream that emerged from meadows above the town. When people in town became convinced that miasmas arising from these newly created wetlands caused several deaths from fever, they insisted that the Rockwells lower their dam and drain the meadows. Since this left them insufficient water to run their forges, the Rockwells moved their fining business to Winsted in 1802.[21]

Winsted had a remarkable concentration of water privileges just the right size for early industries at the outlet of Highland Lake, where water from the lake flowed over a series of falls as it plunged down to the Mad River below. Benjamin Jenkins and his brother-in-law James Boyd, who had brought mechanized scythe making to Winsted in 1792, built a forge on one of these privileges in 1795. The Rockwell brothers, forced out of Colebrook, built their new forge at the privilege upstream of Boyd's in 1803 and added a second forge in 1808, the same year that James Boyd added Winsted's fourth. Ruben Cook built a fifth forge in 1811, on the Still River. Artisans at these works (fig. 5.5) fined pig made by Holley & Coffing at the Lakeville blast furnace. Only the metal made by the most skilled finers met the stringent requirements of the government and private armories for gun iron, the metal used for rifle and musket barrels. The Rockwells and Boyds sold iron of slightly inferior grade for scythes, wire rods, and fine machinery, while a lesser grade answered for blacksmiths.[22]

Willaby Dexter and Joseph Pettee completed a finery forge along with their blast furnace on Mount Riga in 1808. The Holley, Coffing & Pettee partners (see chap. 4 and fig. 4.4) got into the business of making heavy forgings with a second, larger finery forge downstream from the first. They were particularly proud of the anchors made at their forges for the U.S. Navy and for the frigates built in New York in 1821 for the Greeks in their war for independence.[23]

When ordinary grades of bar iron sold for $100 per ton, gun iron fetched a price that averaged $40 per ton higher from 1810 to 1855 (fig. 5.6).[24] High prices attracted other makers into the market for premium-quality metal. Forbes & Adam started fining in addition to bloom smelting in order to make gun iron.[25] In Salisbury, Holley & Coffing at Mount Riga and Canfield & Sterling at the base of the Great Falls made gun iron. Holley & Coffing also built a finery forge adjacent to their new Lime Rock blast furnace. They merged their partnerships into the Salisbury Iron Company in 1828, with Holley serving as president and Coffing as agent at the furnace and forges on Mount Riga.

New Year's Day of 1829 found John M. Holley in New York, frustrated at the continued delay in proof testing anchors his firm shipped to the

Fig. 5.5. Drawing of the last of the finery forges in Winsted. (Connecticut Magazine, vol. 8.)

navy yard. It was cold, and the commandant would allow no fires for heat because of persistent high winds. The same day Alexander Holley was in Springfield with his uncle, Newman Holley, and the son of his father's partner, Charles Coffing, for a two-hour meeting with armory superintendent Roswell Lee. They were to settle the iron company's account and arrange a contract for delivery of gun iron the coming year.[26] A few days later John went on to the nation's capital in Washington, which he described as not much of a place. By the eleventh he had completed the calls

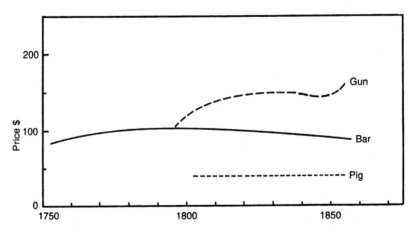

Fig. 5.6. Running average prices of gun, bar, and pig iron in the Salisbury district. The averaging obscures the sharp rise and subsequent fall of prices during the War of 1812. The Civil War caused large price increases, particularly for gun iron. All data are from account books and letters.

needed to get orders for chain cables and continued on to Harpers Ferry to sell gun iron.[27]

John Holley's thoughts about Washington echo those of Peter B. Porter, another son of Salisbury, who visited the nation's capital in 1800. In a letter home he observed, "I have been nearly three weeks at this place and if you have a curiosity to know how I spend my time it is briefly as follows. Conforming myself to the customs of the place I arise at half past seven in the morning, breakfast at nine after which I attend to business when I have any to do, if not, read the news or take a ride around the city. At three I sit down to dinner after which I drink a pint of wine—smoke cigars—talk politicks, & then hear songs. At six I rise from table play a game of billiards and in the evening join a party of girls and boys which is always to be found here, or return to my lodgings & drink & talk & smoke untill twelve, when I go to bed. I love good living. I love the social graces. I love a leisure hour with a friend more than perhaps my northern country man in general, but my soul grows sick of the dayly round of eating, drinking, and playing, and I sigh to return to my native woodlands & the frugal men who inhabit them."[28]

Through 1829 Alexander Holley was often at the iron company's office and store at the Lime Rock forge (fig. 5.7) to supervise the artisans smelting pig iron in the new blast furnace and fining it into gun iron for Springfield and Harpers Ferry. When the blast furnace was going badly, he would call in Ebenezer Brinton (their furnace builder and founder in 1808) to set it right. Furnace management remained an art based on the founder's judgment and ability to read subtle signs. After an episode in which the furnace

Fig. 5.7. Holley & Coffing built this Federal-style combined store and office building at their ironworks in Lime Rock.

ran cold, Alexander learned from Brinton that "When a furnace is losing heat the blaze at the top is thick and whitish. When she is hot or gaining heat the blaze will be clear with blue streaks in it & has a white rim round on the filling plate."[29] In November 1829 he had to deal with a failure at the blast furnace: the lining began to cave in, and the founder prepared on 18 November to "blow her out." Nevertheless, Alexander's duties at Lime Rock left him leisure to enjoy the beauties of Salisbury, as on Friday, 2 October, when he decided to ride to Ivy Mountain, "there to enjoy such a rich treat in viewing the beautiful scenery that surrounds the towering mountain from which the eye sees far over the blue hills, through the long deep vistas of our native land." His diary reflects the growing awareness of the environment and quality of life in Salisbury among his neighbors. Domestic life focused on visiting friends and spending evenings with the family at the fireside reading or around the piano.

John Holley traveled again to Harpers Ferry in 1831. In September he took the newly constructed, horse-drawn B&O Railroad from Baltimore to Ellicotts Mills, covering nearly fourteen miles in an hour and twenty minutes. He rode in a car carrying about thirty people and thought it "a delightful mode of conveyance . . . a more complete triumph of science over nature than even canals." On 22 September he had seen Benjamin Moor, the master armorer, and John H. Hall, manager of the rifle works,[30] who assured him that Salisbury iron was the best they had ever had. (Nevertheless, they had rejected some that Holley & Coffing had sent to them.) While waiting to see Superintendent Rust (then away at his father's place in Loudon County), John described Harpers Ferry in letters home to his wife, Mary Ann. It was a dismal and very sickly place. He preferred his home country with "its habits and improvements" better than anything else.[31]

In 1808 Luther Holley & Son had supplied the crew at the Lakeville blast furnace with as much rum as they wanted. By 1829 the Holleys felt differently about liquor. On Thursday, 15 October, Alexander took eight or ten men "to the Lime Rock Forge to assist in raising a Coal House, 70 x 30 feet, which was done without the use of spirits, except good cheerful spirits of some 150 or 200 of our citizens, refreshed with a plenty of cake, cheese and cider. The building, though large, heavy and difficult to raise was put up without any accident."[32] The Holleys helped found the Salisbury Temperance Society on 4 July 1831; by April 1833 it had 374 members pledged to total abstinence. However, the society still had work to do: it reported that Salisbury then had fifty-six habitual drunkards.[33]

Holley & Coffing made a pitch to increase their defense business with a proposal to Secretary of War Lewis Cass that the National Cannon Foundry then under consideration by the government be built in Salisbury. The partners claimed that the Salisbury ore was the best in the United States and, probably, the world for making cannon, basing their claim on their success in supplying gun iron to the government and private armories. They thought the ore bed might be purchased from its owners for $30,000 to $50,000, and offered to sell one or two of their furnaces to the government. These, with the addition of an air furnace to melt more iron, would suffice for casting the largest cannon wanted by the army. Completion of the proposed railroad from New York to Albany by way of Salisbury would make the means of transportation to the cannon foundry "as perfect as they can be." In fact, the federal government never built a national foundry, here or elsewhere. Nevertheless, government business remained a very important component of business for the Salisbury ironmakers. In 1834 J. M. Holley wrote to his brother Myron that "We are suffering here very severely under the Jackson system of proscription for opinions sake, and the want of an appropriation in Congress, for the expenses of the War department. The ordnance department are indebted many thousands of Dollars to persons in this town for Iron rec'd at the U.S. armories for which no payment will be received untill the apprpriation bill shall be passed." In the 1830s Holley & Coffing's reputation was at its zenith as a supplier to private buyers. Thus, when the Black Rock foundry in New York wanted heavy forgings for a large steam engine it was building in 1832 it turned to Holley & Coffing.[34]

The final addition to the gun-iron makers came in 1835, when Silas B. Moore joined Horace Landon and his brother Albert in Landon, Moore & Company to build the finery forge on Riga Brook in Salisbury Center later run by S. B. Moore & Company. The new firm supplied iron products to nearly two dozen different buyers in Connecticut, Massachusetts, and New York between 1837 and 1841 (some of these were jobbers, who then distributed products to other customers). It dealt with customers as far away as Wilmington, North Carolina, while its principal customer outside New England was the government armory at Harpers Ferry, Virginia.

The largest number of orders to Landon, Moore & Company were for wagon and carriage axles (including axle arms and axle tree drafts). Gun iron sold to the Springfield, Harpers Ferry, and private armories was the second-largest component of the forge's business. Landon and Moore also supplied metal forged to shape specified by customers, sometimes by matching patterns supplied, but more often working to numerical dimensions. Customers usually specified these to the nearest one-sixteenth inch, while the most demanding, such as the government armories, would call for "3/8 scant" or "1/16 strong." Working to such tolerances with tilt hammers challenged the skills of even the best hammermen, and they did not always succeed: among letters received from dissatisfied customers, complaints about failure to hold to specified dimensions were the most common.[35]

Landon, Moore dispatched its products on wagons except when there was good snow cover for sledding. Deliveries to points beyond Springfield went by the Western Railroad. The firm also used the Farmington canal for deliveries to New Haven but sometimes found its products delayed because ruptured banks drained the canal and stranded the boats. Iron for Harpers Ferry went by coasting schooner from New Haven or a Hudson River port to Baltimore or Georgetown, and thence onward by canal or railroad. Buyers had to order well in advance of their needs. Surprisingly,

Landon, Moore received relatively few complaints about delivery problems from their customers.

The Landon, Moore & Company correspondence shows the continued use of barter in commercial transactions through 1840. One customer paid for crowbars with nails; another offered general merchandise in exchange for iron. (Like most ironmakers, Landon, Moore & Company had a store dealing in general merchandise.) Ripley and Cure in Hartford received iron bars and sent back cast and blister steel imported from England, as well as firebricks needed to repair the forge hearths.

Salisbury finery-forge artisans made iron with seventeenth-century techniques. Individual craftsmen who bought iron for use in their own shops could work around variations in the properties of metal made by these old methods. However, the managers of the new manufacturing industries wanted large amounts of uniform metal that would go through their production lines without causing trouble. Quality problems began to plague the gun-iron makers. Into the 1820s Springfield Armory Superintendent Roswell Lee, along with the managers of the Harpers Ferry and the private armories that made arms for the government, had been concerned about the high price of Salisbury iron. By 1829 Lee and the other managers were encountering episodes of high failure rates of musket barrels in proof testing that they ascribed to batches of poorly made iron received from the Salisbury district forges.

Lee, fed up with problems caused by nonuniform iron supplied by the Salisbury makers, sent inspectors to test the product at the forges. The inspectors broke the end of each bar made and examined the appearance of the fracture to judge the quality of the metal. Holley & Coffing, the largest maker, had the highest rejection rate, while the Rockwell and Cook forges in Winsted had the lowest in the 1829 tests (table 5.1). While some of the armory's problems may have been the result of bad welds rather than bad iron, there is little doubt that the Salisbury artisans had been unable to maintain the uniformity of their product as they increased the production rate.

Since they knew little metallurgy and relied on craft skills of hired artisans, Holley & Coffing had difficulty responding to this new challenge. When there was no improvement in the product by 1831, Lee suspended deliveries.[36] Nevertheless, the government armories failed to find a better source of gun iron, and they continued to depend on the Salisbury makers

Table 5.1. Gun Iron Test Results

	Bars broken	Rejected	%
Holley & Coffing	940	102	11
Holley & Coffing second forge	1351	130	10
Canfield & Sterling	248	16	6
Solomon Rockwell	579	54	9
R. Cook	220	9	0.4

Source: Walkley to Lee, 28 December 1829, Springfield Armory records.

for another twenty-five years. The events in 1829–1831 marked the start of changes that would eventually see new entrepreneurs replace both Salisbury's traditional forge business and its leading families as ironmasters.

In the meantime, the district's reputation remained high. While preparing to make his first revolvers, Samuel Colt traveled to learn about suppliers of materials. The superintendent at Springfield told him in 1836 to go to Salisbury for "the best iron made in the United States." He visited Landon and Moore, where Moore assured him of willingness to extend his forge to any degree to meet requirements that Colt might have.[37] At this time Landon, Moore & Company was a major supplier to the national armories. In 1837 the firm sent more than 4,000 barrel plates to Harpers Ferry alone, over half the barrel iron used there.[38] Quality problems persisted, however. A cluster of letters arriving with complaints about wrong dimensions received in June and July 1839 suggest that one or more of the firm's skilled hammermen had left.[39] Also, some batches of bad iron left the forge: J&E North of Berlin complained, "A great part of the iron is hard . . . is worth no more to us than Russian iron." Seth Peck wrote, "We find it impossible to convince any of your forgemen that we want good iron." Asa Waters, proprietor of the Millbury, Massachusetts, armory, asserted that the iron in one lot of barrel blanks "works dry, . . . drops to pieces in the fire." Nevertheless, he badly needed more iron and asked that it be sent right away.[40] In a sample of 200 letters received from 1837 through 1841, there were twenty complaints about products supplied, seven concerning iron quality and the rest failure to hold to specified dimensions. More letters (10 out of the 200) were pleas for the dispatch of more iron, which buyers urgently needed. Thus, meeting the demand for their product was a more pressing problem for Landon, Moore & Company at that time than lapses in quality.

Others challenged the Landon, Moore firm's hold on the market. In 1844 Oliver Ames, the Massachusetts shovel maker, wrote asking for all the iron Moore's works could send but saying that he would not pay any increase in price. He had found that he could get equally good metal from a bloomery forge in the Adirondack region of New York (where low-phosphorus ore was available) and from Norway and Sweden.[41]

Gun iron continued to be a profitable product for the Salisbury ironmakers through the 1840s. In 1845 Canfield & Robbins made about 400 tons of iron of all kinds and about 150 tons of gun iron for Springfield, Harpers Ferry, and Whitneyville. Nearly a third of all the iron the firm made went to the armories. Since this metal sold at a premium price, it contributed nearly half of Canfield & Robbins's income. By 1849, armory sales had dropped to 11 percent of the firm's output.[42] Quality problems increased with increased production in the 1850s. In a sample of fifty-three letters to S. B. Moore & Company (successor to Landon, Moore) from 1850 to 1853, 23 percent complained of bad metal (compared with 8 percent of letters in 1837–1842). W. R. Smith & Company said the iron it received was "flawy and split on punching"; Lyman Wilcox wrote that the "last lot was full of cracks and seams"; Peck & Smith had "poor iron [that] peeled up and finished off bad"; Harpers Ferry noted "much irregularity on quality of iron for musket barrels" and, later, "that supplied has generally been of inferior quality"; and the Springfield Armory said "measures must be taken to ensure better quality."[43] Superintendent Whitney reported that between 1850 and 1855, five-sixths of the barrels made at the Springfield Armory were of Salisbury iron.[44] The armories' business was important enough to S. B. Moore in 1854 for him to offer to make good excess losses of barrels at Springfield, aside from rifling defects.[45]

The quality of gun iron depended primarily on the size and distribution of slag particles within the metal, and on its carbon and phosphorus content. Finery artisans controlled slag and carbon in the metal by their skill in manipulating the formation of the loup in the forge hearth, and in subsequent hammering of the loup.[46] The requisite technique depended on experience and care in executing the necessary manipulations. Iron with excess slag or carbon might be made by inexperienced finers, failure of managers to adequately reward finers for exercise of the requisite skills, or slackness in inspecting the finished product. These were all problems that a Salisbury forge owner could deal with. Control of phosphorus, which embrittled iron, was a more difficult problem.[47]

During the period in which gun iron was an important Salisbury product, most American artisans remained unaware that small amounts of phosphorus embrittled iron, and they lacked any method of determining its presence. They judged the ductility of bar iron from the appearance of the fracture when a bar was broken. Ironmakers in the Adirondacks succeeded in making highly ductile iron by selecting ores available in their region that happened to have low phosphorus contents. No Salisbury ore could match the purity of the Adirondack ores. Whereas finers could remove phosphorus, they did not describe what they did in this way. They could only guess at the amount of this impurity in the pig they started with, or how much they removed during fining. As Salisbury miners dug deeper into their ore deposits to meet the demands of increased production, they brought up ore with varying phosphorus contents. The smelters' and finers' craft-based techniques could not cope adequately with this problem, which worsened as they attempted to increase production of gun iron.

Frustrated by their problems with welded-iron barrels, private arms makers such as Whitney and Remington substituted barrels made of cast steel imported from Sheffield, England. In 1858 the Springfield Armory switched to barrels made of English iron welded in rolls. Had the Salisbury ironmakers invested in the research needed to make this iron, they would have had an enormous market for it during the Civil War.[48] Instead, they let the forge branch of their business wither at a time when the price buyers would pay for their product was increasing.

SIX

Environment, Technology, and Community in Salisbury

Salisbury's particular combination of natural resources allowed iron-making to continue for nearly 200 years from its 1736 start. In the first hundred years of mining and smelting, the district's entrepreneurs had established a place for themselves in the national market for high-quality iron products while they managed their mineral, waterpower, and forest resources to meet the demands of expanded production. Each resource offered different challenges. An ironmaker could cut wood without thought of replacement or could manage woodland for sustained yield. As miners dug deeper for ore, they had more spoil to dispose of. They also had to devise drainage systems for their pits. Waterpower systems needed maintenance, repairs after freshets, control of their watersheds if siltation of the power ponds was to be avoided, and cooperative agreements among users. To meet the demands of expanding business, a forge owner might have to develop a new power site, as Forbes & Adam did when they put their second rolling mill in Woodville instead of Canaan.

Fuel supply eventually proved the most critical natural resource problem for the ironmakers. Woodland previously burned by Indians made a good initial source of ironworks fuel, since wood from small trees coaled easily. The Lakeville blast furnace had a capacity of 2.5 tons of iron per day and fuel consumption of 250 bushels of charcoal per ton of iron made.[1] In 1776 the furnace operated seven months, the length of a typical blast, and so would have made about 525 tons of iron while consuming fuel from about 200 acres (0.3 square miles) of woodland. At this rate of production, the furnace would have consumed at most 5.4 square miles of forest, 9 percent of the area of the town of Salisbury, between its start-up date (1762) and 1780.

An eighteenth-century bloomery typically made about 250 pounds of iron per day with a fuel rate of about six. If it worked 300 days per year (a high estimate), it would have used wood at the rate of 42 acres per year. From 1736 to 1780, one bloomery would have consumed the wood from 2.9 square miles of forest. This estimate is probably high, since ore had to be brought to the early bloomeries in saddlebags over trails. The sixteen bloomeries built before 1780 (see table A1.3) would have used the wood from 46 square miles of forest in towns with a total area of 438 square miles. The woodland burned by Indians, along with the uncut forests, could have supplied ironmakers with all the fuel they needed through colonial times, leaving plenty to meet other demands, such as home heating, sugar boiling, and lime burning.

Although no one had to worry about a wood shortage in colonial times, the cost of gathering fuel did increase as forge and furnace proprietors consumed woodland adjacent to their works. Transportation costs from increasingly distant woodland entered Richard Smith's decision to put up his finery forge in Colebrook rather than Lakeville. John Cotton Smith, in the report he prepared for the Connecticut Academy of Arts and Sciences, noted that three of the five forges in Sharon had closed by 1800 because of the increased cost of fuel, even though the rugged hillsides would supply all the wood anyone would ever want.[2] Nevertheless, fuel prices advanced steadily (fig. 6.1), even when the general price index was not rising.

To deal with the increased demand for fuel, both the merchant capitalists and the artisan-entrepreneurs had to organize assured sources of charcoal. They could either buy woodland and hire cutters and colliers to make their

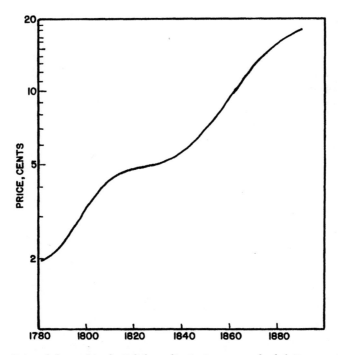

Fig. 6.1. Price of charcoal in the Salisbury district in cents per bushel. Data are from account books and ledgers kept in the district.

charcoal or buy coal from independent contractors. An ironmaker who owned woodland could manage it for continuous production. Proprietors began to buy up large acreage in the 1790s. Forbes & Adam, which acquired extensive woodlands to supply fuel to its forges in East Canaan, contracted with artisans such as Elijah Pettebone to cut and coal the wood. Pettebone's November 1796 contract for 20,000 bushels of "cole" (charcoal) specified that the partners would allow him to cut 400 cords of wood on their land (or more if needed), would pay him 2 shillings per cord, and would retain title to the wood. Next spring, summer, and fall Pettebone would coal the whole and deliver it to the forge. The partners would pay as the coal was delivered at 20 shillings per load of 100 bushels (after deducting sums advanced for cutting the wood), regardless of whether the price rose or fell, for half the coal; for the other half, they would pay the going price at the time of delivery.[3] They used both hard- and soft-wood charcoal in increasing quantities as their business expanded (table 6.1).

Holley & Coffing's purchases show the amount of woodland that ironmakers needed to secure a reliable fuel supply. When Luther Holley acquired the Lakeville blast furnace (in 1799), he also got 1,800 acres of forest with it. The furnace, running near capacity, consumed the wood from 200 acres a year. Since it took on average about twenty years for a new crop of trees suitable for coaling to mature, Holley would have needed 4,000 acres of woodland in continuous production for a sustained fuel supply. He had to more than double his woodland to keep the furnace running at capacity.

By 1812 Holley & Coffing had two furnaces and a forge in operation. They needed nearly 10,000 acres of woodland and then had only 2,300 acres, leaving them further from having a sustainable fuel supply than they had been in 1799. In 1829 they had three furnaces and six forges in operation, and made 1,000 tons of pig iron and 450 tons of bar iron.[4] They would have used 5,630 cords for the furnaces and 4,561 cords for the forges. (These figures agree well with their reported wood consumption of 10,000 cords.)[5] They would have had to cut about 500 acres of woodland a year.

By 1819 the woodland Holley & Son began cutting in 1799 would have been ready to yield another crop of fuel. The second cutting reduced the amount of new woodland Holley & Coffing had to cut by 224 acres a year; after 1828, twenty years after the Mount Riga furnace went into blast, they would have been cutting second growth from 448 acres a year. Improve-

Table 6.1. Forbes & Adam Charcoal Purchases

	Oak	Chestnut	Pine	Total
1790–1794	2,717	103	1,958	4,778
1794–1797	4,693	759	1,991	7,443
1797–1799	9,249	1,359	3,835	14,443

Source: Data from Kenneth Howell and Einar W. Carlson, *Men of Iron: Forbes and Adam*, Lakeville, Conn.: Pocketknife Press, 1980, p. 101.
Note: Quantities are in average bushels per month.

ments in fuel efficiency and management of woodland for sustained yield would have brought them close to a continuously renewable fuel supply. To achieve this they would have needed about 8,500 acres of woodland committed to growing wood for coaling. They then had about 7,000 acres. All the ironworks in the town of Salisbury would have needed about 12,000 acres committed to charcoal production in 1827 to sustain their existing production rates, and an additional 230 acres per year to meet increasing demand.

Everyone in the town of Salisbury heated their homes and did their cooking with wood fuel. A household used about thirty cords per year.[6] Salisbury had about 400 households at that time, consuming the wood from 600 acres each year. The town would have needed 12,000 acres for woodland to assure its domestic fuel supply. Townspeople would have used additional wood for activities such as making maple syrup and burning lime. Altogether at least 25,000 acres (thirty-nine square miles), nearly half the area of the town, would have been committed to growing fuel wood. This land would have been a patchwork, with some plots freshly clear-cut and others wooded with trees with trunks up to about six inches in diameter.

When nearby woodland could not meet demand, an ironworks proprietor could import wood (or charcoal) from a distance, paying the increased transportation costs. While wagon haulage was expensive, the Housatonic River gave the region one avenue for bringing in wood cheaply. Horatio Ames took advantage of this route when he began fuel-wood drives down the Housatonic shortly after opening his ironworks at the Great Falls in 1832 (see chap. 7).

The waterpower that drove the machinery at all the ironworks also taxed managers' skills. Large natural ponds supplied the Lakeville and Mount Riga furnaces. These and the three furnaces that drew from the Housatonic always had adequate water available. The other forges and furnaces depended on smaller streams, and drought could leave them short of water at times. Managers had to allow for seasonal variations in water availability in planning their work. Where streams had several users, forge and furnace owners had to work out cooperative agreements on the management of water releases, such as those at the Fuller forge in East Kent (see chap. 3).[7] Conversion of woodland to pasture or tillage increased soil erosion. In upland settings, the eroded soil collected near its source in the smaller streams. Power ponds on these streams were more likely silted up than those on the larger rivers.[8] A forge or furnace owner could avoid costly maintenance and prolong the life of his power system if he undertook management of the watershed above his works.

. Everyone could see the heavy pressure on woodland and the need to use water prudently in the Salisbury district. However, no one visualized anything other than an inexhaustible supply of ore. Ore Hill, half owned by the Livingstons of New York, remained the region's largest mine. Ironmakers could also get ore from the Chapin, Chatfield, Davis, Indian, Porter, Star, and Scoville pits. The mine at South Kent supplied forges in Kent and neighboring towns and the Kent blast furnace. The proprietors of the Macedonia furnace imported ore from nearby mines in New York. Furnacemen could shift to a new source if miners encountered difficulties at a particular mine. They could get as much limestone for flux as they wanted from nearby pits throughout the region.

Easily dug ore supplied the forges and furnaces throughout colonial times. Miners did not need sophisticated mining techniques, since the ore occurred in pods embedded in clay, from which they could easily extract it. They dug their own pits and loaded the ore in carts that horses would then draw up inclines to the surface. By the time of Benjamin Silliman's 1817 visit, miners at Ore Hill worked in a number of large, deep excavations. As they dug away overburden to get at the ore, they piled up mounds of waste adjacent to their pits. By the 1780s miners at the Kent ore bed had to sink shafts forty to eighty feet deep to reach ore. Many were injured as they blasted the ore loose and hauled to the surface with windlasses. About 1790 the mine proprietors agreed to pay Phillip Judd, the neighboring land-owner, fifteen dollars a year for water from his brook to wash away over-burden. Work in the open pit mine was safer and easier. To get permission to run a drainage ditch through the land that their neighbors Jackson Bull and Sherman Boardman owned below the mine, the Kent proprietors had to agree to keep the ditch clear and not permit mine waste to run over the meadows.[9] The area where their ditch discharged is a boggy wetland known today as Mud Pond (fig. 6.2). The Kent mine that Silliman visited in 1818 was one big pit 150 feet deep and several hundred feet wide.[10]

Holley & Coffing, who had started with a tenth interest in Ore Hill, in 1812 acquired the Davis ore bed to avoid the high royalty charged by the other Ore Hill proprietors. They found that the Davis ore required much washing to separate it from its accompanying clay.[11] The clay-laden waste water must have been a considerable nuisance as it flowed from the wash-eries into streams near the mines.

If anyone in Salisbury considered the use of woodland for fuel produc-tion or deposits of mine wastes objectionable, they left us no record of their thoughts. Small trees still mantled most of the hillsides, much as they had in 1736, revealing open vistas now gone. The mines were suffi-ciently far from each other that they made isolated islands of disturbance in the lightly forested landscape. Plenty of streams well stocked with fish still ran through the countryside.

One factor that helped reduce the amount of industrial waste strewn about was that both iron and its slag lent themselves to recycling. A smith could reshape any piece of scrap in his forge and could weld small pieces together if he needed larger ones. At the large forges, such as Hunts', arti-sans placed old iron in the bundles they hammered into anchors, shafts, and other large items. Because of this recycling, it is rare to find remnants of iron at forge sites today. Slitting-mill proprietors bought scrap to in-clude in the billets they heated for rolling. The Forbes & Adam account book for their Washington mill records scrap iron, scale from the rolls, and bottoms from the furnace held in stock for sale and recycling.[12]

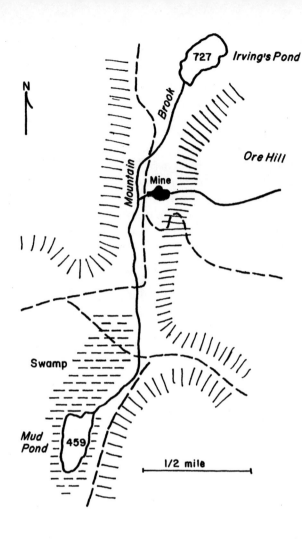

Fig. 6.2. The proprietors of the Kent mine drew water from the adjacent mountain (called Ore Hill) to flush mine spoil into Mountain Brook. The steep slope of the brook moved water past the mine fast enough to carry the spoil a half mile to the level area now known as Mud Pond, where it formed wetland. Elevations are shown in feet.

The volume of slag produced by a blast furnace is about half the volume of the ore charged into the furnace. On his 1817 journey Benjamin Silliman was struck by the piles of slag and refuse around the Lakeville furnace.[13] Today we find little slag at the sites of the early furnaces because ironmasters recycled it. The blast furnace slag contained small drops of iron that were sufficiently valuable to be worth the labor of breaking up the slag with sledgehammers or a stamp mill. Recyclers picked out the iron and dispersed the pulverized slag, which resembled sand, into the surrounding soil or streams. Forge owners in Kent bought slag from the proprietors of the nearby Macedonia furnace to run through their stamp mill, and their finers then converted the recovered shot iron into wrought iron.[14] Since the Mount Riga works had no stamp mill, workers there broke up the slag with sledgehammers to recover the iron it contained.[15]

A finery forge made a volume of slag equal to about 90 percent of the volume of the pig iron it consumed. It contained nearly as much iron as the Salisbury ore. Landon, Moore & Company sent slag from their finery forges to the Hunt bloomery in Canaan and to the Rockwell forge in Winsted for resmelting.[16] Artisans at the Winsted forges reworked their finery slag in bloomery fires to make "a strong, coarse iron suitable for tires, axle and crow-bar patterns, plow moulds, heavy shafting, saw-mill cranks, etc." Each forge had associated with it a drafting shop with lighter hammers to draw down the bars to rods and shapes, and to work up the refuse iron by welding to each piece a layer of blistered steel and drawing the composite into sleigh shoes.[17] Little slag can be found at these forge sites today. Slag and metal wastes, unlike mine spoil, had little impact on the environment in Salisbury through the mid–nineteenth century. However, blast furnace owners would later stop bothering with recovering shot iron and would create large piles of slag near their furnaces.

The work environment in Salisbury ironmaking, while industrial, offered artisans opportunities to master difficult tasks and gain prestige for their skills. It was distinctly different from that in textile mills, where workers tended machines that carried out endless repetitions of the basic mechanical motions of spinning and weaving.[18] Nor did the ironworks employ women, as the paper mills in nearby western Massachusetts did.[19] Ironmakers managed invisible transformations in materials through their ability to read subtle clues from the appearance, feel, and smell of hot metal and slag. The relation of manager and artisan (when not the same individual) was necessarily one of mutual respect. Ironmaking artisans were proud of their work and would move on if they thought themselves ill treated. William Belcher wrote to explain why he was leaving the Ames Ironworks in 1844: "I thought you was dissatisfied with my drawing the iron. therefore it is ever my motto that when I cannot please my employer to leave which I trust will be for the best."[20]

Whether working as inside contractors, as at the Forbes & Adam forge, or running their own enterprises, artisans needed managerial as well as technical skills. Salisbury colliers and forgemen negotiated their own contracts and had control over their work. Holley & Coffing, in a letter responding to Roswell Lee's complaint about high iron prices, noted that forcing down the cost of coaling labor "is difficult to effect in that class of people."[21] Contractor artisans worked at the Forbes & Adam forge until at least 1851.[22] Colliers worked as independent contractors for blast furnace proprietors through the nineteenth century.[23]

Ironmaking artisans needed physical strength and faced possible injury from encounters with hot metal, rushing water, and massive hammers. John Seymour, a worker at the Robertsville forge, drowned when a flume broke in 1775.[24] The Housatonic River swept Walter Holley and David O'Neil over the ninety-foot drop of the Great Falls as they tried to clear the flume at the Ames works in 1837. Holley survived, much bruised; O'Neil's body washed up some miles downriver days later. Two years later Alexander Morris, a fifteen-year-old boy working alongside Horatio Ames, was caught and pulled under a hammer to his death while helping forge hot iron blooms.[25] In Goshen, Frederick Gentile, a French-Canadian collier, fell through the top of a charcoal pit into the fire in 1872. He was fatally burned despite the efforts of his coworker, Jacob Avery, to pull him out. Avery's hands were severely burned.[26] In 1869 "John Burns was killed at the Chatfield Ore Bed in Salisbury when a large chunk of ore, thrown loose by a blast, fell on him. One of his legs was so badly crushed that he died in ten minutes from loss of blood."[27] "On November 27th, Mr. John Danchy, Overseer at the Chatfield Ore Bed, was killed, and a workman seriously injured, when three kegs of powder blew up, while they were approaching to examine a defective fuse."[28]

Artisans coming to Salisbury to work brought knowledge of metallurgical technology used elsewhere. Simeon Palmer, who ran the Wadsworth & Kirby mill in Litchfield, learned rolling and slitting at Taunton.[29] John Adam Jr., Willaby Dexter, Joseph and Seneca Pettee, and later Horatio Ames, John Eddy, and Leonard Kinsley all came from southeastern Massachusetts, bringing expertise to the Salisbury district. John Adam Jr., son of an artisan and an artisan himself, often gathered technical data on furnace and rolling-mill equipment during his travels to see customers.[30]

While an individual artisan could make and sell iron, even individual proprietors found advantages in division of labor. Salisbury artisans and entrepreneurs created a complex network of social and commercial interactions that they recorded in their account books as they organized themselves to utilize the region's resources. Sales on national rather than local markets required managerial tasks difficult for an individual artisan to undertake. In the division of labor at the Forbes & Adam works, artisans negotiated terms for their services; Forbes exercised his role as proprietor in organizing work at the farm, gristmill, and forge; while Adam traveled to deal with distant buyers.

When they undertook to run the Lakeville blast furnace, Holley and Coffing began to develop centralized management through vertical integration, partnerships, and placement of family members to deal with their woodlands, furnaces, forges, mills, stores, and sales of their specialized products. By 1810 they were making 500 to 600 tons of pig iron a year in their Lakeville and Mount Riga furnaces. They extended their vertical integration by building two forges on Mount Riga. In addition to making bar iron, they supplied cast-iron pipe for Albany's first water system. A sampling of Holley & Coffing's orders for 1818 shows them supplying bar iron to Townsend's rolling mill in Albany and Harris's scythe works in Hammertown (north of Salisbury village), gun iron to the Springfield Armory and to Lemuel Pomeroy in Pittsfield, bar iron to Forbes & Adam to be rolled for the Springfield armory, sawmill cranks and screws to a dealer in New York, and pig iron to foundries in Taunton and New York City. They enlarged their capacity in 1825 with an integrated ironworks at Lime Rock that included a blast furnace and finery forge, near the site of Lamb's 1736 bloomery.[31]

One managerial task that Salisbury men rarely undertook was to visit ironworks in other regions to learn about technique or new methods. The extensive Holley family correspondence and the records of the Holley & Coffing partnership show that neither John M. Holley nor John C. Coffing, whose backgrounds were mercantile, brought technical information back to Salisbury even though they corresponded with ironmasters elsewhere about tariff issues. Neither of them knew, or interested themselves in, the metallurgy of ironmaking. Their skills were in organization and sales, and they did not themselves participate in innovation or development of technique. The only significant exception was Horatio Ames, the outsider from southeastern Massachusetts, who undertook visits to ironworks in the Middle Atlantic states and to Europe to observe the techniques used by others (see chap. 7).

In the established culture of Salisbury, distinctions between ironworks' artisans, owners, and managers were blurred. Samuel Forbes, an artisan with managerial skills, took advantage of an expanding industrial economy to acquire wealth and property. Forbes, true to his artisanal background, remained involved in the day-to-day operation of the partners' ironworks, mills, and farms until a few days before his death at age ninety-eight in August 1828.[32] By then he was, according to Timothy Dwight, one of the wealthiest men in the state.[33]

Artisans could still enter ironmaking without the aid of family connections or outside capital well into the nineteenth century. Individual contractors dug ore at the Davis mine in 1835–1838 as they had a hundred years earlier and delivered it to the Forbes blast furnace.[34] Mining offered artisans a way to accumulate capital to enter business. Peter P. Everts, son of a farmer with a common school education and a teamster, began mining on his own account at age twenty-one. By 1848, at age thirty-seven, he was able to buy a share in Ore Hill, and from 1849 to 1871 he served as agent for the proprietors.[35]

William C. Sterling of Lakeville had no college education and was a storekeeper when he joined with Lee Canfield and others in an ironworks venture at the foot of the Great Falls. In 1825 (at the age of thirty-three) he, his brother F. A. Sterling, and others formed Sterling, Chapin & Company (later Sterling, Landon & Company) to run a forge and blast furnace business that continued to 1840. He was an incorporator of the Housatonic Railroad, a bank director, and a partner with A. and S. B. Moore in the satinet factory north of Chapinville. Hiram Weed of Sharon was the very model of an individual entrepreneur. He built a blast furnace small enough to operate all on his own in 1845 and smelted iron with it until at least 1866. Weed, in partnership with J. A. Cray, built the last bloomery forge in the Salisbury district to smelt magnetite from a small deposit found in Sharon.[36]

Charcoal making also offered individuals opportunities to develop artisanal and managerial skills. Harlow P. Harris began has career in charcoal burning at age thirty-three and farmed for two years before entering the iron business. He became superintendent of the Richmond blast furnace when he was thirty-seven years old, then of charcoal kilns in Vermont; in 1871 he was a railroad foreman, and in 1872 superintendent of the Chatfield Mining Company's operations, where he remained until his death in 1885. Harris knew a lot about Salisbury coaling practice, but all his experience was local, and despite numerous changes of job, he remained uninformed about professional developments elsewhere.[37]

Family continuity kept successive generations of the Holley family in ironmaking. Luther Holley's son, named John Milton Holley, joined the family mercantile and ironmaking business in 1799. John's three daughters married ironmasters.[38] The son John and Mary Ann named, significantly, Alexander Hamilton Holley grew up in comfortable circumstances, inherited money from his grandfather, traveled, and often remarked on the natural beauties of the places he visited.[39] He became a clerk for Holley & Coffing at age sixteen and a partner in Holley & Company at twenty-one. His diary for 1829 shows that he spent much time traveling about the region, visiting friends and relatives, in addition to supervising furnaces and woodlots. He traveled to customers such as the Springfield Armory to collect bills or negotiate contracts (see chap. 5). Later he participated in state politics, served as governor of Connecticut, and promoted railroads and banks. He brought a knife works to Salisbury, one of the district's few manufacturing enterprises.[40]

Family connections played a major role in the Barnum-Richardson Company, the firm that eventually presided over the decline of Salisbury ironmaking. Milo Barnum came to Lime Rock from Dover, New York, about 1820 and with his son-in-law, Leonard Richardson, acquired control of the foundry in Lime Rock. A few years later Barnum made his son, William H. Barnum, a partner in the firm that eventually controlled all the region's furnaces. Successive generations of Barnums and Richardsons ran the business up to its end in the twentieth century.

Salisbury owners, managers, and artisans lived in the same communities, and in proximity to the ironworks, until the industry was virtually defunct. Both J. M. and A. H. Holley admired their region's natural beauties. At a time when ironmaking heavily used Salisbury's natural resources, A.H. Holley could note "the beautiful scenery that surrounds the towering mountain from which the eye sees far over the blue hills, through long deep of our native land."[41] Everyone in the iron business also farmed, some, like J. M. Holley, on a large scale. Holley also traded in real estate, so that at the time of his retirement in 1834 he had made himself Lakeville's principal property owner. His homestead included a large farm. People recognized Horatio Ames as a good farmer. Residents saw industry and farming as compatible uses of the region's resources. Unlike in the Middle Atlantic states and Virginia, where ironmasters organized plantations, individual artisans in Salisbury carried on their own farming enterprises.

Farming and other enterprises kept the economies of most district towns diversified. In Canaan in 1846, agricultural products made up nearly as large a component of the town's economy as the value of the iron made by its forges and furnaces (table 6.2). Of the district's towns, Kent was the most heavily dependent on the iron industry.

Table 6.2. Components of the Economy of Canaan in 1846

	$	%
Agricultural products (less hay)	81,385	29
Forge	71,790	25
Forest	45,644	16
Textiles	31,360	11
Blast furnace	27,700	10
Miscellaneous manufactures	25,162	9

Source: Daniel P. Tyler, *Statistics of the Condition and Products of Certain Branches of Industry in Connecticut,* Hartford: Boswell, 1846.
Note: Hay is not included in agriculture, since most of it would have been converted to animal products.

Young people in Salisbury grew up in a community that placed a high value on education. Thirty-four residents subscribed amounts ranging from £1 to £5 to form a Library Association in November 1771. Richard Smith also contributed, and had nearly 200 books—most on divinity, philosophy, and history—sent from London to form the initial collection. In 1803 Caleb Bingham added 150 volumes. The records of the association, with their numerous fines for books returned with grease on their pages (presumably from the candles used by night-time readers), show that Salisbury residents really did use their library.[42] Peter Porter, who grew up in Salisbury, was surprised to find during a stay in the nation's capital in 1800 how few people in Washington read books. In a letter to his friend John M. Holley he remarked, "The people here with few exceptions never fatigue themselves with the inspired amusement of reading and among thirty boarders who lodge at the James Hotel with me I was plagued to find a book to cheat away the hours." Even entrepreneurial ironmakers in Salisbury read extensively: John Adam Jr. had a library of 174 volumes at the time of his death. However, it contained but one scientific book, *Conversations on Chemistry.* [43]

Formation of the Bachelors' Club by a group of young men in 1827 indicates the intellectual climate attained in Salisbury by the early nineteenth century. The members included A. H. Holley and his brothers, as well as William P. Burrall, later president of the Housatonic and the Illinois Central Railroads and husband of Alexander Holley's sister, Harriet. They wrote and discussed original essays that they printed in their journal, the *Bachelors' Repository.*[44] The Holley family letters describe a home life of evenings around the piano or reading. They record a civil and well-ordered family life coupled with strong interest in the temperance movement. The anonymous author of the 1888 A. H. Holley memorial volume remarked on the integrity of the people in the area at the time of his business, "the best period of New England life," referring to a kind of golden age.

Commitment to education was one of New England's strengths as the colonies and, later, the early Republic moved into industrialization. The colonial government required each town (or, after 1712, each society) to provide a public school and contributed to its support from its revenues. The General Assembly applied much of the proceeds from the auctions of the Western Lands to school support. Next to hiring a minister, school budgets and personnel were usually the most important business at society meetings. Salisbury engaged a teacher in 1743, established school districts in 1745, and built a schoolhouse in 1747.[45]

John Milton Holley attended local school to age eleven, boarded in Boston to learn English and penmanship, and went to an academy in Williamstown and then Williams College for a year. In Salisbury he learned surveying from Samuel Moore, author of *An Accurate System of Surveying*, published in 1796. In 1796 John went to Ohio to survey new lands for the Connecticut Land Company and then spent two years managing his father's business interests in Dover, New York. Although he did not complete college, he cultivated literary interests and accumulated an extensive library that contained religious and philosophical tracts, an encyclopedia set in the process of completion, but no books on science or technology.[46]

In the early nineteenth century many Salisbury families sent their sons out of town for their schooling, and the most ambitious expected matriculation at Yale to follow. John Adam Jr. sent S. F. Adam to Yale in 1799. A. H. Holley attended academies in Sheffield, Hudson, and South Litchfield up to age sixteen, but avoided Yale because of ill health. J. C. Coffing sent his son John Henry to military academy at Norwich, Vermont, at age thirteen and Stockbridge Academy at fifteen, and had him start learning the iron business as a clerk for Holley & Coffing at age eighteen. His second son, Churchill Coffing, completed a master of arts degree and law school at Yale. The third son, George C. Coffing, went to public school, the Salisbury Academy, and an academy at Argyle, New York. George subsequently had a successful business career as manager of the Richmond, Massachusetts, furnace in 1853, an officer in several other furnace firms, and part owner of textile mills. He ran for Congress and indulged his interest in fancy cattle. He had varied interests rather than the focus of an iron professional.[47]

A. H. Holley, intending that his son A. L. Holley complete a classical education at Yale, sent him to the appropriate preparatory schools. However, A. L. Holley's interests ran strongly to engineering. At age eighteen he completed an essay on cutlery that reveals substantial knowledge of metallurgy and manufacturing and a capacity for expository writing. He detested learning Latin and Greek, and after a struggle convinced his father to send him to Brown, which had just started a course of study in engineering.[48] A. L. Holley seems to have been the only Salisbury son to aspire to an engineering education; he never returned to apply it in his home district.

The prominent place of artisans' skills and liberal education in Salisbury shaped attitudes toward science, innovation, and the environment. In 1736 natural science had little to contribute to artisans' understanding of metallurgical technology. Forge or furnace proprietors attended to finance and sales, and relying on artisans to make and shape the iron with craft knowledge they learned from their elders or through experience. A hundred years later, chemists had learned how to describe the basic iron-making processes and could ascribe some properties of iron, such as brittleness at high or low temperature, to the presence of specific impurities, such as sulfur or phosphorus. Salisbury forge and furnace proprietors continued to attend to finance and sales, and artisans to rely on their craft skills. This left all of them ill prepared to deal with their customers' increasingly stringent demands for higher quality and uniformity in the metal they bought at the premium prices the district charged, and threatened the particular social relationships that shaped the Salisbury ironmaking community.

The lack of sophistication in technical and managerial skills in the district led to the failure of an ambitious engineering project, the Water Power Company in Falls Village. In 1845 Lee Canfield and Samuel Robbins, successful proprietors of a furnace and two forges, thought they had the necessary resources and expertise to utilize the Housatonic's power potential at the Great Falls. With the aim of creating another Holyoke or Lowell, they organized the Water Power Company with a capital of $200,000. (Family connections were important in this enterprise: Robbins and W. P Burrall, a director, were sons-in-law of J. M. Holley.) The company dug a channel at the top of the falls to divert water into a 2,100-foot-long upper-level canal. Factories were to draw water from this canal and discharge it into a parallel, second-level canal 30 feet below. Additional factories were then to use the 30-foot drop to the third-level canal and a final 30-foot drop to river level, thereby realizing the full power potential available at the falls in stages that existing waterwheels could easily handle. The power company built its three canals with massive, rubble stone walls that stand today. A popular story asserts that when, in 1851, the proprietors first let water into their system, the canals leaked so badly they flooded the adjacent town fairgrounds and made nearby roads impassible with mud. Although canal builders often encountered leakage problems when building on limestone, successive retellers have undoubtedly enlarged the difficulties experienced by the power company.

After Canfield died in 1859, William H. Barnum bought his interests. Samuel Robbins had close ties with the established Salisbury ironmaking families, while Barnum represented a new generation of entrepreneurs lacking long-standing family connections in Salisbury. Robbins probably found his new partner uncongenial. Despite the high demand for water-power throughout Connecticut during the Civil War years, Barnum and Robbins failed to bring any customers to their canal system. In 1867 the company was cementing the bottom of its canals to combat severe leakage.[49] Although it eventually fixed the leaks, it failed to attract any customers despite the proprietors' intermittent sales efforts over the next forty years.[50] The massive canal walls stand today in Falls Village, a monument to entrepreneurial provincialism.

Although growing up in the business, as either an artisan or a family member, brought personal success to many, this route into management contributed to the eventual decline of Salisbury ironmaking. Lack of engineering education and sparse interaction with colleagues in other districts made it easy for Salisbury's native sons to let their works drift into obsolescence. Their backgrounds left them ill prepared to deal with problems outside their immediate experience, as when Canfield and Robbins undertook their hydraulic power project.

The history of S. B. Moore's Salisbury Center forge shows the limitations of the district's merchant managers in a business that year by year demanded increasing technological sophistication. Like many other Salisbury entrepreneurs, Moore had a common school and academy education. He kept store at Falls Village, at age twenty-seven joined a partnership operating the Chapinsville blast furnace, and in 1832 undertook management of the forge. Like the Holleys and the Coffings, Moore lacked background in metallurgy, had no way of keeping up with technical developments in the field, and had to rely on the artisans he employed for technical expertise. This left his iron enterprise vulnerable to being overtaken by developments outside the Salisbury district. As described in chapter 5, Moore's artisans had increasing difficulty meeting the standards demanded by the armories in the 1850s and matching the quality of iron imported from Norway. Nevertheless, Moore and his partners did well in the forge business, being worth about $100,000 in 1857. With profits in hand and faced with new technological challenges, he decided not rebuild his forge after it burned in an 1859 fire. Moore subsequently served as town treasurer and probate judge.[51]

In 1855 forge owner Lee Canfield, the heirs of John C. Coffing and John M. Holley, and several other Salisbury families collected a dividend of $4,050 from the Davis Ore Bed Company, the profit on $5,793 of sales.[52] Generous royalties like these from property acquired by their parents left the sons of many established Salisbury families with little incentive to keep up with the iron trade. Others such as William Adam sold off family-owned land at a good profit. (Adam sold the land for the village of Canaan once the Housatonic Railroad arrived.)

At the time that Moore, the Holleys, Coffings, Canfield, and the Adams were taking their profits and dropping out of the iron business, opportunities remained open to them in the high-end market for charcoal-hearth iron and crucible steel. Over in Collinsville, just outside the Salisbury district, the entrepreneurial managers of the Collins Company began steel-making in 1865. Just thirty miles away, brass makers prospered in Waterbury, a location with no natural-resource advantages, because of the knowledge of the community's skilled artisans and the entrepreneurial vigor of its managers. Salisbury had the skilled artisans but, faced with technological change that would call for a different use of its land, chose a different course.

SEVEN

The Challenge of New Markets
and Techniques

Ironmakers in the Middle Atlantic states used canals and railways to reduce costs and expand the scale of production with new techniques
based on mineral-coal fuel beginning in the 1820s. Salisbury forge and furnace proprietors, who still had teamsters hauling ore, fuel, and metal
along dirt roads with wagons in summer and sleds in winter, knew that
improved transportation systems would help them get their products to
outside buyers. They were less aware that canals and railroads would
eventually force them to confront new techniques adopted by ironmakers
outside their district.

Entrepreneurs in northwestern Connecticut had become interested in
waterways as early as 1760, when they wanted to improve the Housatonic's channel north to Massachusetts in order to float logs downriver to
their sawmills. Although the General Assembly authorized a lottery to
raise £300 for the project in 1761, the promoters accomplished nothing.[1]

The start of construction on the Erie Canal stimulated interest in building a canal along the Housatonic River that would open new markets for
the northwest's ironmakers. Urged on by John M. Holley and others, the
Ousatonic Canal proprietors organized a company in 1822 to build from
tidewater to Stockbridge, Massachusetts. However, when canal engineer
Benjamin Wright's survey showed the company would have to build
enough locks to raise boats a total of 604 feet as they traversed the canal,
the project's supporters backed out. The promoters of the Sharon Canal
project, intended to start in Sharon and go down the Oblong River into
New York and thence follow the route later used by the Harlem Railroad,
accomplished even less.[2]

John M. Holley had experienced railroad travel on his 1831 trip to
Harpers Ferry. He and his neighbors realized that a railway up the
Housatonic valley would gather traffic from the region's ironworks and,
with a connection to the Western Railroad in Massachusetts (fig. 7.1),
open the first year-round route from New York City to Albany. (The railroad along the Hudson River between New York and Albany did not open
until 1851.) Several of the region's ironmasters, including J. M. Holley's son
A. H. Holley, helped raise funds for the construction of the Housatonic
Railroad when the state issued a charter in 1836. Salisbury ironmaster

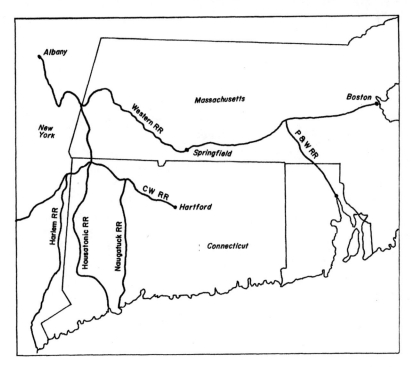

Fig. 7.1. Map of rail routes serving the Salisbury district. The Housatonic Railroad reached the district first, running from Bridgeport on the coast to a connection with the Western Railroad. The Naugatuck and the Harlem offered adjacent routes to the coast. The Connecticut Western (CWRR), the last to be built, met the Naugatuck at Winsted and crossed the Housatonic at Canaan. The Providence & Worcester (P&WRR) met the Western at Worcester.

William C. Sterling served as an incorporator, and William P. Burrall was railroad president. The company's wooden rails covered with iron straps started in Bridgeport on the coast, reached New Milford in 1840, and Sheffield, Massachusetts, with a connection with the Western Railroad in 1842. In the early years of operation, snake heads (iron torn loose from the rails) caused frequent accidents. The company solved this problem by rebuilding its track with solid iron rail in 1846.[3]

As soon as the year the Housatonic Railroad opened, Salisbury ironmakers began using it to ship out their products. Thus in 1842 Canfield & Robbins sent bar iron and forged products by rail to Bridgeport, where the railroad company forwarded them to customers in New York City by either sloop or steamer. Others, like Horace Landon, tried to dicker with the company about its freight rates: in December 1842 Landon asserted that unless he got a rate reduction he would send his pig iron from Chapinville by sleigh instead of rail. By 1846 S. B. Moore & Company used the railroad to send iron bars made at its forge to the Ames Manufacturing Company in Massachusetts and to Lyman Wilcox in Berlin, Connecticut, as well as gun iron to the Springfield Armory. It received files, shovels, and other hardware items it purchased in Bridgeport by rail.[4] In addition to aiding merchants and ironmakers, the Housatonic line led communities to refocus their town centers on the railroad: buildings in West Cornwall put up after the arrival of the railroad faced the tracks (fig. 7.2). People in Canaan moved the center of their town to Falls Village, a community they built around the depot. In North Canaan, residents created a new village, named Canaan, built along the rail line.

Parallel lines on the east and west sides of the Salisbury district further improved its links with outside markets. Just over the state line in New York, the builders of the Harlem Railroad had reached only as far north as Dover Plains by 1849. The railway company then hired A. H. Holley to get subscriptions for an extension of the track to Millerton, which it reached in 1851. To the east, the backers of the Naugatuck Railroad surveyed their route in 1848 and constructed the line with iron "H" rail in 1849.[5] It opened service to the forges and other industries concentrated in Winsted. Connecticut's terrain made east-west construction difficult, and Salisbury had to wait until 1871 for a line that could deliver ore to the Canaan blast furnaces from Ore Hill (see chap. 8).

While almost anyone could see the value of railways to Salisbury, the new ironmaking techniques adopted in the Middle Atlantic states presented the district's ironmakers with a more difficult challenge. The first of the new methods, puddling, offered an alternative to fining and was the most economical means of making the large quantities of wrought iron Americans needed for rails, bridges, and machinery. British opinion held that only long-flame bituminous coal would work in a puddling furnace. The coal available in eastern Pennsylvania, where the demand for iron was concentrated, was anthracite (which burned with a very short flame). By 1830 Pennsylvania artisans had designed a new hearth that allowed

Fig. 7.2. The Housatonic Railroad passes through the center of West Cornwall today, with the village's nineteenth-century buildings facing the tracks.

puddling furnaces to work with anthracite, and they soon had large-scale production under way: the Montour works in Danville had twenty-two puddling furnaces when its artisans rolled the first American "T" rail in 1845; by mid-century they had made the Montour the largest rail mill in North America. Virginia entrepreneurs used the James River and Kanawha Canal to bring pig iron from furnaces in the Shenandoah valley and bituminous coal from Virginia mines to Richmond, where they had waterpower and a seaport. By 1837, Virginians had puddling and rolling mills, including the famous Tredegar works, selling bar iron to customers in seaports along the east coast.[6]

In the second of the new techniques they adopted, Pennsylvanians replaced charcoal fuel in their blast furnaces with anthracite. To do this they had to add more powerful stoves to preheat the air pumped into their furnaces, and it took a decade of experimentation to get the process working reliably. They had an anthracite-fired blast furnace producing pig iron in 1838, built another one in 1839, and built five more in 1840.[7]

Pennsylvanians needed additional years to master the technique of producing uniformly good-quality metal with mineral-coal fuel. Then, even after metallurgists discovered how to use anthracite to make iron with the same properties as charcoal-made metal, some buyers, suspicious of this innovation, would pay a high price for the traditional product. (Samuel Collins, proprietor of the great Connecticut axe works, retained an abiding, and unjustified, distrust of any iron made with mineral-coal fuel.[8]) Ironmasters in the Middle Atlantic states continued to use charcoal fuel to make the best grades of bar iron. To use this expensive fuel as efficiently as possible and simultaneously make better metal, they brought the old techniques of bloom smelting and fining to new levels of sophistication. New Jersey forgemasters developed the American bloomery process, which remained in service into the beginning of the twentieth century in the Adirondack region, making iron for crucible steelworks (see chap. 3). Pennsylvania ironmasters developed the seventeenth-century finery process into the charcoal-hearth technique for making pure iron with a minimum amount of carbon and included slag.[9] It also remained in use through the nineteenth century. Salisbury ironmasters ignored both of these innovations.

As the entrepreneurship of the Holley, Canfield, Adam, Hunt, and other Salisbury ironmaking families waned, three relatively new arrivals in Salisbury—Leonard Richardson, William Barnum, and Horatio Ames— took their places in guiding the region's ironmaking. Instead of making small quantities of the finest grade of wrought iron, they wanted to sell large tonnages of products such as car wheels and locomotive tires to railroads.

In 1830 Oliver Ames, the eastern Massachusetts shovel maker, had capitalized on the demand for tools among the contractors building the Erie Canal by sending his second son, Horatio, along with John Eddy, to Albany to sell tools. Strong demand for iron products led Horatio, along with Eddy and Leonard Kinsley, a member of an established family of Massachusetts ironmakers, to enter the forge business. In 1832 they bought land from the

Adam family at the top of the Great Falls of the Housatonic (fig. 7.3) for a new ironworks. The high reputation of Salisbury pig iron coupled with an immense waterpower attracted Ames and his partners to this location. However, they had no canal to bring in mineral coal. To compete with forges elsewhere, they had to use the puddling process. They solved their fuel problem with the first innovation in metallurgical technique made by anyone in the Salisbury district, redesigning the firebox of the conventional puddling furnace to burn wood fuel (fig. 7.4). Then, to get a good supply of fuel wood, they ran log drives down the Housatonic from land they bought in Massachusetts. They also applied the great waterpower of the Housatonic to drive the helve hammers artisans used to make products such as crowbars and wagon axles.

By 1835 the Ames family had bought out Eddy and Kinsley, and had made Horatio manager of the forge. He brought new entrepreneurial vigor, sometimes untempered by sound judgment, to the enterprise. In addition to being an innovator, Horatio often worked alongside his artisans at their furnaces and hammers. He shared their risks, as when he navigated a rowboat across the brink of the falls to get a log boom in place.[10] He and many of the artisans lived adjacent to the forge. Puddling iron demanded skill, experience, and physical strength. Good puddlers were aristocrats of labor, commanded respect in the community, and expressed their standing in their Amesville houses (fig. 7.5). Nearby fields allowed them to farm as well as practice their metallurgical arts.

Fig. 7.3. This historic photograph shows the Great Falls of the Housatonic River before a modern hydroelectric plant diverted much of river's flow to its turbines. Spray from the falls has covered the tree on the left with ice. The Ames works is just out of sight behind this tree. (Courtesy of the Falls Village–Canaan Historical Society.)

Fig. 7.4. *In this view of the interior of the Ames ironworks, a puddler is grasping a ball of white-hot metal with tongs suspended from an overhead track in order to carry it to the water-powered helve hammer across the aisle. Molten slag drips from the metal; residual slag will be forced out as the metal is consolidated under the hammer.*

Fig. 7.5. *This house owned by one of the puddlers at the Ames works stands with similar homes, all basically unaltered, on Puddlers Lane in Amesville today.*

Once Ames showed the way, the Salisbury Iron Company and others in the northwest began to adopt the wood-fired puddling technique he had pioneered (see table A1.4). The Kent Iron Company in Macedonia built one of the largest puddling works. Here twenty-five men made iron aided by teamsters, and other workers made all kinds of wrought-iron work, such as crowbars and wagon tires, sent in great quantities to Poughkeepsie, New Haven, and Bridgeport.[11] Like most Salisbury ironmasters, Eber S. Peters, the overseer, followed by his son J. H. Peters, ran a nearby store and sawmill. They prospered, and built a substantial stone house opposite the mill. (This handsome house is now incorporated into an estate.)

Railroad building created new opportunities for Salisbury's ironmakers well before the first train rolled into their region. Americans had started building steam locomotives in the 1830s and needed heavy iron forgings for their axles. Ames had his ironworks making these by 1835, gaining a substantial market share even though teamsters had to haul the axles to Hudson River ports for shipment to his customers. In 1840 he added tire bars to his line of railroad products.

Until about 1830 anyone around Salisbury who wanted a casting had to get it poured at one of the area's blast furnaces. Furnace campaigns lasted only a few months, and if no furnace were in blast, a customer might have to wait months for his casting. About 1830 someone started a small foundry in Lime Rock where artisans could remelt pig iron kept in stock to pour castings whenever a customer wanted one. Milo Barnum, a Lime Rock storekeeper, and his partner Leonard Richardson got control of the Lime Rock foundry and carried on its business in a small way in connection with their store, making clock and sash weights, plow castings, and so on. About 1840, they began making chairs, frogs, and other track fittings for the Western Railroad, then being built from Springfield to Albany. They shipped their castings on wagons hauled by teams to the railroad at Chatham, New York, about fifty miles away. A few years later their artisans began casting chilled-iron car wheels, which proved to have excellent wear resistance and helped enlarge the reputation of Salisbury iron. Nevertheless, in 1846 artisans in the town of Salisbury produced only 170 tons of castings while making 2,000 tons of pig iron and converting most of it into 1,800 tons of forged products.[12]

The Adam heirs had not fared as well as some of the other district families. Charles S., Forbes S., and George Adam, grandsons of John Adam Jr., owned the old anchor forge, the unmodernized Forbes blast furnace built by S. F. Adam, the family gristmill, and much land in East Canaan. Short of cash in 1855, they formed a partnership with Richardson, Barnum, John Beckley, and William Pierce for the manufacture of pig iron and the milling of grain. The Adams retained the right to raise their forge dam (but not so much as to interfere with the furnace wheel) and draw water to "flow the meadows." The Richardson group would put up all the cash to fix up and run the furnace and the grist mill, and would split the profits equally with the Adams.[13] From this beginning, Barnum and Richardson would later go on to gain control of all the district's furnaces (see chap. 8).

Horatio Ames bought all the pig iron for his forge from nearby blast furnaces. From 1836 through 1838 he received deliveries of two to three tons of iron per day from the Adam family's Forbes furnace, a substantial part of its output.[14] The strong market for pig iron created by Ames's forge and the other new puddling works led Salisbury ironmasters to invest in new furnaces and make improvements to old ones, once the region recovered from the lingering effects of the panic of 1837 (see table A1.1).

Many owners of charcoal-fueled blast furnaces outside of the Salisbury district promptly adopted improvements in technique developed for furnaces that burned mineral coal. William Henry's 1835 experiments with a recuperative hot-blast stove at the Oxford furnace in New Jersey showed how fuel consumption could be lowered by 40 percent.[15] Within five years many Pennsylvania and Maryland ironmasters had installed hot blast stoves in their furnaces.

Holley & Coffing had been complaining as early as 1820 about the expense of charcoal fuel and the cost of bringing it from increasingly distant sources.[16] Nevertheless, they approached the use of hotblast stoves at their furnaces very cautiously. One reason was that some customers, such as the army ordnance officers responsible for procuring cannon, continued to insist as late as the 1850s that iron made with hot blast was inferior. However, the Salisbury ironmasters did not sell much iron for cannon founding. Instead, through the 1830s and 1840s finery forges and puddling works converted most of the region's pig into bar iron. Once a founder learned how to manage a furnace fitted with hot blast, he could make forge pig as easily as in a cold-blast furnace and substantially cut his fuel costs: converting to hot blast could reduce the fuel rate from 1.6 to 1.1.[17]

No Connecticut ironmaster adopted hot blast until the owners of the Bulls Bridge furnace in Kent added a stove in 1844. Others followed cautiously over the next two decades (see table A1.1). However, the managers of the Salisbury Iron Company left their Mount Riga furnace, used exclusively for forge pig, without hot blast when they rebuilt it in 1846. The Adam heirs had not put hot blast on the Forbes blast furnace and were still suspicious of its value in 1855: when they terminated the lease of the furnace to the Hunt-Lyman Company, they required that Hunt-Lyman remove the hot-blast apparatus it had added.[18] Some ironmasters elsewhere in the United States continued to run ancient cold-blast furnaces. Nevertheless, the tardy adoption of the new technique in Salisbury is another indication of how much the district's furnace owners had distanced themselves from metallurgical innovation.[19]

While the opening of the Housatonic Railroad offered new opportunities, it also forced Salisbury blast furnace owners to face another new challenge. The railroad could deliver mineral coal, a fuel Pennsylvanians ironmakers adopted, and which their Salisbury counterparts had been able to ignore while it was out of their reach. After 1842 they had to make decisions about coal-based ironmaking technique. By then, Pennsylvanians already had fifteen years of experience making wrought iron in puddling furnaces, and they were bringing coal-fired blast furnaces rapidly into production. Both techniques made iron at lower cost than the wood- and charcoal-fired Salisbury furnaces.

Ironmasters in Port Henry, New York, and West Stockbridge, Massachusetts, north of Connecticut, demonstrated that they could smelt iron profitably at furnaces near their ore mines with anthracite shipped in from Pennsylvania by canal or railway. Proprietors of the Pomeroy Iron Works in West Stockbridge built their anthracite-fired blast furnace in 1850 to run on fuel delivered by the Housatonic Railroad. They ran it successfully and rebuilt it in 1872 to keep it in production. They could make 10,000 tons of pig a year. By 1848 Horatio Ames had Pictou, Nova Scotia, and Cimberland, Maryland, coal delivered by ship to Bridgeport and then brought to Falls Village by rail. These bituminous coals worked well in his puddling furnaces, and allowed him to abandon the dangerous and difficult log drives down the Housatonic River that he had previously used to get furnace fuel. Construction of a railroad spur on a bridge over the river allowed him to receive coal deliveries at the shop door (fig 7.6).

The owners of the Kent and Macedonia blast furnaces took a gradual approach to the problem posed by mineral coal. After they rebuilt their furnaces with hot blast in 1846 (Kent) and 1847 (Macedonia), they made additions of up to 50 percent anthracite to their charcoal fuel. (Pennsylvanians had discovered that it was much easier to run a furnace with mixed fuel than with pure anthracite, which required a higher blast temperature than charcoal-fired furnaces.) The results they obtained with the mixed

Fig. 7.6. By 1850 the Ames works had its own railroad siding crossing the river on a bridge (out of sight to the left) from the Housatonic's main line visible in the foreground. In the background are houses occupied by Ames and his artisans. Beyond are the fields where they raised vegetables and livestock. A flume brought water from the dam to the turbines that drove the forge hammers, lathes, and other machinery.

fuel were successful enough to encourage Camp and Fenn, owners of the neighboring Bulls Bridge furnace, to rebuild so that they could use pure anthracite fuel. Operating as the Monitor Ironworks, they remained in production until 1865, even though they were over two miles from the railroad. The proprietors of the Macedonia furnace abandoned smelting in 1860, probably because of their remote location. They had a nearly three-mile, uphill wagon haul from the Kent depot to bring in mineral fuel (fig. 7.7). The railroad could deliver anthracite directly to the Kent furnace, which remained in blast until 1892. Other Salisbury district furnace proprietors apparently never tried mixed fuel in their furnaces.

In addition to puddling furnaces and water-powered helve hammers and machine shops, Horatio Ames installed large steam hammers (developed in England by James Nasmyth in 1839) at his Great Falls ironworks. He ordered the first, one of the largest Nasmyth had yet made, on his 1849 trip to England. Ames claimed, probably correctly, that it was at that time the largest in the United States. Artisans used it to make locomotive axles,

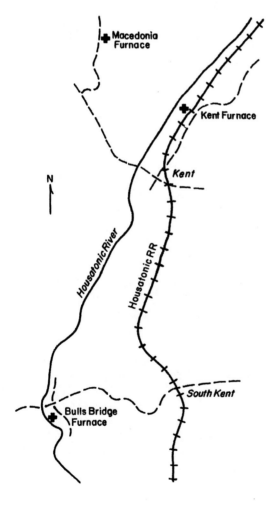

Fig. 7.7. The managers of all three blast furnaces in Kent experimented with mineral-coal fuel delivered by the Housatonic Railroad. Teamsters had to climb 200 feet in the 2.9-mile haul to the Macedonia furnace from Kent depot. Reaching Bulls Bridge from South Kent involved a climb of 296 feet from the depot, followed by a 363-foot descent along the 2.4-mile route to the furnace. Kent furnace had its own siding.

massive steamboat propeller shafts, and other heavy forgings (fig. 7.8). In 1855 Ames put in new equipment that made his works an effective competitor in both price and quality to the English works that had held a tight grip on the American locomotive tire market (fig. 7.9). By then the Ames family had invested $400,000 in plant and equipment that engaged 200 artisans when in full operation. Ames's artisans could forge a locomotive axle and cut its throws in a day, heat and draw fifty car axles a day, and make 5,000 locomotive tires a year, enough to satisfy half of the United States demand.[20]

While Ames went ahead with new techniques, his neighbors, Lee Canfield and Samuel S. Robbins, at their works at the base of the Great Falls, made ineffective attempts to adapt to new markets and methods. Gun iron had been an important part of their business until they failed to maintain the necessary quality. Sales of bar iron, forged car axles, head blocks, and bumpers to the Housatonic Railroad had been an increasingly important component of their business from 1844 through 1855. The firm began receiving small shipments of coal in 1844.[21] In the late 1850s they made railcar axles and used both charcoal and mineral coal. Ironmakers had known since the seventeenth century that they could not substitute mineral coal for charcoal in a bloomery or finery hearth because the direct contact between the metal and fuel at a high temperature resulted in contamination of the metal with the sulfur present in the coal. Neverthe-

Fig. 7.8. Teams of artisans forged ships' propeller shafts and other massive iron products from red-hot metal under the steam hammer they named "Thor" at the Ames ironworks.

Fig. 7.9. Horatio Ames built this specialized machinery to form iron rings into perfectly round tires for locomotive wheels.

less, artifacts collected at the forge site show that Canfield & Robbins tried mineral coal in their fining hearths. The resulting metal was contaminated with iron sulfides and phosphides, and would have been worthless. Attempts to save money with mineral coal may have caused Canfield & Robbins's difficulties in satisfying the Springfield Armory,'s quality standards for gun iron. Artifacts also show that Canfield & Robbins had adopted the Pennsylvania technique to the extent of at least one coal-fired puddling furnace, using the iron to forge railroad axles.[22]

Canfield and Robbins's involvement with their unsuccessful Water Power Company venture along with their ineffective attempts to modernize their forge, left them ill prepared to weather the panic of 1857, and they abandoned their forge the next year. The Salisbury Iron Company, the Boyds, and the Rockwells had all closed their finery forges in the early 1850s. People tried to explain away this lack of enterprise in terms of uncontrollable, external factors. Thus, according to Winsted's historian, that town's forges closed by 1853 because of the government's adoption of gun iron from Norway (in fact, untrue), competition from English makers who offered better products at lower prices, adoption of puddling in "more favored localities," advances in the price of wood and charcoal, and the high cost of transportation.[23] As we have seen, these were not the real reasons.

Ames survived the depression financially weakened. To recoup, he undertook manufacture of large, wrought-iron cannon that attracted national attention during the Civil War. His works made at least six fifty-pounders (later bored to eighty-pounders) and fifteen seven-inch guns.

Government tests showed the Ames guns to be superior artillery, although they cost nearly a dollar per pound, about ten times the cost of cast-iron cannon. Ordnance officers, who did not believe the guns were worth the extra cost, delayed paying Ames for them. Horatio, never known for nice manners or refined language, grew increasingly testy as his relations with government officials deteriorated; he got deeper and deeper into debt until losses from the cannon project left his works bankrupt. Less than two months after Horatio's death in 1871, his brother, Oliver Ames Jr., sold the property to the Housatonic Railroad. The railroad sold the iron on hand in the forge for $50,000, thereby recovering much of its $75,000 purchase price. The largest steam hammer, know as "Thor," went to the American Silver Steel Works in Bridgeport, and the railroad built its roundhouse on the site of the main forge shop. By 1880 all the ironworkers had left, replaced by railroad employees, thus ending over a hundred years of Salisbury expertise in iron forging.[24]

Civil War contracts induced at least one other Salisbury entrepreneur besides Ames to go into ordnance production. Andrew Hotchkiss of Sharon patented a novel explosive shell for rifled cannon in 1860. The pointed shell had a detonator activated by impact. The Hotchkiss company, founded by Andrew's father to make hardware, devoted the entire capacity of its Sharon foundry to shells in 1862, and in 1863 it moved production to Bridgeport.[25] Hotchkiss family members later set up large ordnance works in Europe.

British sympathy with the Southern cause and the depredations of the Confederate navy during the Civil War emphasized to Americans their dependence on imported steel at the very time Europeans were experimenting with new techniques for large-scale steel production. A group of Hartford investors led on by expert, but as it turned out inaccurate, advice from Benjamin Silliman and other Yale professors undertook the district's only venture with new steelmaking technique. Through the mid–nineteenth century, Europeans used a variant of the puddling process to make steel instead of wrought iron. The Hartford investors, who lacked experience with either metallurgy or the Salisbury district's natural resources, accepted Silliman's assertion that the vein of ore in Roxbury's Mine Hill was particularly suited for making steel. By 1867 they had a mine opened and a blast furnace and steel puddling works built. The German artisans they hired to puddle steel from the pig made by the blast furnace found themselves unable to reproduce their European process in Roxbury, and the whole enterprise soon foundered. The Mine Hill venture—with its absentee owners, imported artisans, and reliance on scientific advice—was alien to every aspect of established Salisbury practice. The physical remains, now in the care of the Roxbury Land Trust, the most extensive and complete in the district, are an impressive monument to failure.[26]

Rolling mills offered ironmakers a new method of shaping bars and rods, and the possibility of making plates and sheets larger and thinner than could be made with hammers. Bloomers and finers had traditionally used helve hammers to work the excess slag out of their metal and trip hammers to finish the bars. A hammerman needed considerable skill to make a bar with some pretense of uniformity. We saw in chapter 5 how customers increasingly complained about poor hammer work at Moore's forge. By the 1830s Pennsylvanians had learned to build substantial rolling mills that could shape iron bars much faster and more precisely than could artisans working with tilt hammers. Within a few years of their Pennsylvania competitors, ironworks proprietors in southern Connecticut began to use squeezing machines and rolling mills in place of hammers to make wrought iron: the Greenwich Iron Works had four roll trains operating in 1836, and the Birmingham works in Derby put in four trains in 1847.[27]

After Samuel Forbes and John Adam Jr. completed a rolling and slitting mill in East Canaan in the 1780s similar to the one built at Saugus 140 years earlier, Salisbury's entrepreneurs largely ignored developments in rolling technique. Horatio Ames, the Salisbury district's innovator, designed, patented, and built a rolling mill in 1847 that twisted an iron bar after its first hammering so as to work its slag fibers into a spiral pattern. He aimed to improve the mechanical properties of the iron by reducing its tendency to split along the straight fibers left by hammering.[28] Ames's mill did not shape the iron into a finished product. Like his neighbors, Ames relied on hammering for this; as he had in 1844, he continued to lament that his hammers were too heavy to make shovel plate and that his attempts at it ruined his equipment.[29] Others had plenty of trouble hammering out the thin plate increasingly in demand. Thus, in 1843 Landon, Moore & Company had to make excuses for the poor quality of the shovel plate they hammered out for Oliver Ames's works in Massachusetts.[30] Rolls would have solved these problems.

A partial explanation for the reluctance of the Salisbury ironmasters to adopt rolling mills is found in doubts some customers continued to harbor about the quality of rolled iron. Back in 1825, Springfield Armory superintendent Roswell Lee had advanced the claim that hammering removed slag more effectively from bar iron than did rolling, even though he had been instructing Holley & Coffing since 1818 to send iron to the Forbes & Adam mill for rolling into plate and strip.[31] Careless artisans or unscrupulous ironmasters could make bad iron with either technique, and in metallurgy old prejudices that may have once had some validity died hard.

Intense demand for iron plate at the Springfield Armory as it expanded production in 1861 created an opportunity for William J. Canfield, president of the Hunt-Canfield Iron Company of Huntsville, to finally bring rolling mills to the Salisbury district. William's father, Lee Canfield, the forge proprietor and banker, had left him money to invest in a new venture. William, along with Edward P. Hunt, converted the old Hunt forge into a rolling mill to make the plate the armory used for gun locks. They had one puddling and two coal-fired heating furnaces, two water-powered tilt hammers, one nine-inch roll train, and three breast wheels to provide power.

After the war, Hunt-Canfield shifted to making tires, railroad and carriage axles, and merchant bar.[32] In June 1867 the works had just finished an order for 1,400 axles for the Chicago & North-Western Railroad, working night and day since September, and were at work on 260 axles for a railroad in Cuba.[33] The next year Hunt-Canfield made axles for the Union Pacific Railroad. After a fire at the mill in 1869, Canfield decided not to rebuild, largely because the old Hunt forge site lacked adequate waterpower. Canfield could have moved his works to a site on the Water Power Company's canal system at the Great Falls, where he also could have had a direct rail connection. Instead, he gave up.

Other Salisbury ironmasters and manufacturers shared Canfield's lack of entrepreneurial vigor. A correspondent wrote in the *Connecticut Western News* of the need for enterprise in Salisbury. He observed that the district's resources of waterpower and iron were not fully used.[34] A Norfolk correspondent reported that the town was nearly deserted in 1872, whereas twenty years earlier it had been "a smart business place."[35]

The closure of the Ames works and the Hunt-Canfield forge left the Barnum-Richardson Company the sole survivor in the Salisbury district. Only sales of pig iron to customers who still believed in the value of the Salisbury name, along with railroad car wheels, remained for the district's ironmakers.

EIGHT

Retreat from Progress

Salisbury ironmakers throve by selling wrought iron rather then cast iron through the first half of the nineteenth century. Their finery forges and puddling works converted nearly all of the pig produced by the district's furnaces to bar iron or forged products. However, by the 1860s, when the district's ironmasters were smelting up to 11,800 tons of pig iron per year, they converted little of it to wrought iron. The demise of the forges left just one principal product, cast iron used mainly for railroad car wheels. Milo Barnum and Leonard Richardson had started making railroad castings in 1840 (see chap. 7). When Milo Barnum retired in 1852, his son W. H. Barnum took his place in the partnership with Richardson. The partners expanded the business by acquiring the Beckley and Forbes furnaces in 1858 and 1862, respectively, from the Adam family in East Canaan. Upon Leonard Richardson's death, Barnum and the Richardson heirs reconstituted the business as the Barnum-Richardson Company, the firm that gradually gained control of all mines and blast furnaces in the northwest, except for the Kent furnace.

A new railway facilitated the Barnum-Richardson operations. Dedicated residents of the northwest, in the face of much skepticism, raised the capital needed to build the Connecticut Western Railroad from Hartford to State Line, where it joined with the Dutchess & Columbia line running to Beacon, New York (see fig. 7.1). Salisbury residents eagerly awaited its 1871 completion: they wanted to be rid of the heavy ore wagons that kept their roads a mess passing from Ore Hill to the furnaces in East Canaan. The Connecticut Western passed through Winsted, traversed difficult terrain in Norfolk, and crossed the Housatonic Railroad at Canaan, where the two companies built a handsome union station (fig. 8.1).[1]

Railroad enthusiasm also led residents in the northwest to propose impractical schemes. The Shepaug Railroad had been completed in 1872 from Danbury to Litchfield. A correspondent writing to the *Connecticut Western News* that year proposed extension into the Salisbury district. He noted that water privileges were less important than they had been, that only iron manufacture set the district's towns apart from other impoverished Litchfield County farm towns, and that the Housatonic Railroad had caused a large increase in the iron business. A railroad through War-

Fig. 8.1. The Canaan union station stands at the crossing of the Housatonic Railroad, whose track is in the foreground, and the Connecticut Western. The agent, in his office at the corner of the building under the tower, sold tickets and controlled the signals that guarded the crossing.

ren, Huntsville, and other towns would tap much wood and iron freight. Since the Shepaug line had never earned a return for its backers, and the extension would be through difficult terrain, few shared the letter writer's enthusiasm. In fact, the district had seen the last extension of its rail network.

Barnum and Richardson's car wheel business throve after the Civil War. The company built a second foundry in Lime Rock in 1870 and added another blast furnace in East Canaan in 1872. By 1877 it had acquired seven furnaces: three in East Canaan and one each in Lime Rock, Sharon Valley, Cornwall Bridge, and Huntsville (see fig. A1.1). It built a new wheel foundry in Chicago in 1873, the first venture of a Salisbury firm outside the district. Its foundrymen could cast 300 wheels per day in Chicago and 200 in Lime Rock, all made of Salisbury iron (fig. 8.2). An unrelated foundry in Jersey City with a capacity of 150 wheels per day used iron from the Buena Vista furnace.[2]

Barnum-Richardson's vertical integration did not yet include the iron mines. Three pits—Ore Hill, Chatfield, and Davis—produced most of the ore in the Salisbury region. Miners no longer worked on their own as independent contractors but instead worked for mining companies. Two companies worked the pits at Ore Hill and paid royalties to the mine proprietors. The largest, the Brook Pit Company, had fifty men and twenty horses (fig. 8.3) raising 150 tons per day and expected to get 25,000 tons out in 1872. Miners took the ore to crushers and passed it on to washers. Here streams of water carried the lumps through rotating drums made of perforated cast-iron plates and washed off the adherent clay. The water then took the ore lumps to a second drum with narrow slits. Workers returned lumps that failed to fall through the slits to the crusher. The Connecticut furnaces used about 45,000 tons of the 55,000 tons of ore dug in 1872; the rest went to nearby furnaces in New York.[3] By 1880 it cost the mining companies $1.75 for labor to get a ton of ore at a time when the national average was $1.35.[4] The eight Connecticut blast furnaces (Sharon, Kent, Chapinville, Cornwall Bridge, Beckley, Buena Vista, Lime Rock, and Canaan No. 3) then had a combined capacity of 91 tons per day.[5]

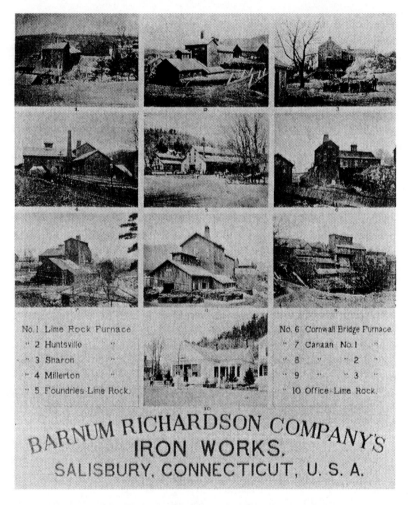

Fig. 8.2. The Barnum-Richardson Company advertised itself with this poster showing its office and foundry in Lime Rock, as well as the blast furnaces it controlled. Company managers may have felt that the antiquated appearance of the furnaces bespoke the ancient reputation of Salisbury iron on which it traded. (Photo from the C. R. Harte collection, courtesy of Fred Chesson.)

Fig. 8.3. Miners at the Ore Hill open pit used two-wheel, horse-drawn carts to remove overburden and haul out ore in the 1870s. (Photo from the C. R. Harte collection, courtesy of Fred Chesson.)

While industries elsewhere turned to mechanization to ease manual labor, work at the Salisbury blast furnaces remained little changed from what it had been for decades. At the Cornwall Bridge furnace in 1853, the staff worked alternate twelve-hour watches. Individuals performed a range of tasks, as required by the exigencies of furnace operation. There was no job specialization, except for the founder, and the company paid a day rate that was almost the same for all kinds of furnace work.[6] Coaling, however, remained the province of individual artisan-contractors. Each collier contracted to make and deliver a share of the fuel the furnace used.[7]

At the Buena Vista furnace, 200 men and 150 horses were employed making ten tons of pig iron daily.[8] Horse-drawn carts delivered ore and fuel (fig. 8.4). Fillers charged the furnace from hand carts while laborers prepared the sand pig bed. After the founder or keeper tapped the furnace, workers broke the pigs free and stacked them outside the casting house, awaiting teamsters to haul them to a depot or to the Lime Rock foundry.

The region's blast furnaces achieved their maximum output in 1882, when they smelted over 24,000 tons of iron. Although most of the iron went into wheels, the Kent Iron Company sold its pig to customers, such as Schenectady Locomotive Works and the Harris-Corliss Engine Company, that still believed in the advantages of charcoal-smelted iron for machinery components. Agriculture remained a component of the iron business at Kent. The company continued to grow grain on its fifty acres of fields and operate its gristmill to provide food for teams of horses that hauled ore, flux, and fuel to the furnace.[9]

The prosperity of the 1880s masked the technological backwardness of the Salisbury ironmakers. In 1857 the Pioneer Iron Company called Stephen R. Gay of Salisbury, who had learned charcoal blast furnace technique at the Cornwall Bridge furnace, to Michigan to build the first furnace to smelt Lake Superior ore.[10] Fifteen years later no one would have turned to Salisbury for such help. While the Barnum-Richardson Company built its East Canaan furnace in 1872 to the same plan Salisbury artisans had used thirty years earlier, Alabama ironmasters, who also made railroad car wheels, completed the charcoal-fired Tecumseh furnace with the latest improvements used at the Pennsylvania anthracite-fired furnaces: cupola construction, closed hearth, bell-and-hopper charging, steam-powered blowing, and Player-type stoves.[11] Although the Salisbury industry had yet to reach its peak output, and would then linger for another forty years, its management had been dormant with respect to technique since 1863.

Consolidation of the remaining furnaces under control of the Barnum and Richardson families in the 1870s intensified managerial stagnation in

Fig. 8.4. In this view of the Buena Vista blast furnace, the casting house is in the foreground, while the furnace is enclosed within the tall central building. Behind are the enclosed charging bridge and one of the sheds used for storage of ore and charcoal. The man standing on the bridge over the Hollenbeck River (the source of power for the furnace's blowing engine) appears to be a supervisor. Teamsters delivering ore and fuel have assembled in the rear for the photographer. (Photo from the C. R. Harte collection, courtesy of Fred Chesson.)

the district. Furnace manager James Bierce's letters to Oliver Ames asking for advice on the operation of the Cornwall Bridge blast furnace illustrate the problem: William H. Barnum took over as president of the furnace company in 1869. He also had a career in national politics that began with his election to Congress in 1867, continued with appointment to the Senate in 1876, and included service as Democratic national chairman in 1880 and 1884. His outside interests distanced him from the day-to-day operation of his iron company, and by 1874 Bierce was complaining that Barnum had not been to the furnace for over two years. Bierce had to turn to Oliver Ames, an outside stockholder, for help.[12]

The members of the American Institute of Mining Engineers visited Salisbury in 1879. The report that Rossiter Raymond, the institute's secretary, wrote of the meeting extolled the hospitality of the Holleys and other prominent families, whose daughters had helped entertain the visitors. Raymond asked one young engineer, "How is it that you, who devoted fifteen minutes to an ore bed and ten to a blast furnace, found it necessary to spend a full hour in the study of a young lady?" The members greatly enjoyed an encounter with Governor Holley's fast-trotting ox team, which, they said, passed everything on the road. They felt the region had much to attract the mining engineer and the lover of nature, but one member, noting the obsolete furnaces, wondered if "we would ever again visit these delightful places." He perceived that Salisbury had already begun its transition from an industrial region to new uses of its land and resources.[13]

Five years later the members of the United States Association of Charcoal Iron Workers enjoyed an excursion to Salisbury as part of their sixth annual meeting, held in New York City. The members commanded enough interest for the Housatonic and the Connecticut Western Railroads to provide a private train for their trip, and Salisbury residents to take them on excursions in their carriages. The editor of the association's journal, John Birkinbine, was impressed with the beauty of the country, which included rugged hills, fertile valleys, pleasant lakes, and imposing cascades. In the face of such hospitality, Birkinbine could hardly be overly critical. Nevertheless, he remarked that Salisbury colliers still made charcoal largely in pits (other regions had kilns) and transported it in wagons (railcars were used elsewhere). He noted the crude, antiquated mining appliances: the miners still dug ore in open pits with decades-old methods using only hand drilling; they cleared away dirt to get at the pods of ore, which they broke up by blasting where necessary, and hauled out in two-horse dump carts to Blake stone crushers. Birkinbine noted the "ancient form of blowing apparatus" used at the furnaces and politely reported that "to most of the visitors the furnace structures and appliances did not appear modern." One of the visitors, H. M. Day, was less inhibited, observing that "to many the appliances for mining and smelting appear quite crude and antiquated." Both writers remarked on the beauty of the scenery: the "rugged hills, and fertile valleys . . . beautiful lakes and imposing cascades, lent a charm to the trip which cannot be conveyed by the pen."[14] Yet while the mines and furnaces drifted deeper into obsolescence, the Barnums, Richardsons, and their junior partners were using the profits from their ancient furnaces and obsolete mines to build themselves new mansions in Lime Rock, the village that housed their company's home office. These included C. W. Barnum's Hephzivalla (fig. 8.5), M. B. Richardson's Edgewood, and R. N. Barnum's Foxhurst, which stood in ostentatious contrast to the houses of the established Salisbury families.

Another indication of decay was the under-representation of Salisbury in the membership of the United States Association of Charcoal Iron Workers, the group that met frequently and published a journal devoted to the technology of their industry. In 1885 Salisbury accounted for 5 of the association's 317 members.[15] The district then produced about 3 percent of the nation's charcoal iron output, and its proportional representation would have been 11 members.

The district's antiquated furnaces could continue to operate only because Salisbury pig sold for as much as 60 percent more than some other varieties of charcoal pig. The ironworkers association's editor suspected (rightly) that this advanced price depended more on an ancient reputation than on any actual superiority. Alexander Lyman Holley, as knowledgeable as any American about ferrous metallurgy at that time, noted that the reasons that some irons chilled (as required to make a hard rim on a car wheel) and others did not in particular applications remained unexplained.[16] Consequently, foundrymen relied on experience with particular brands of iron to get the results they wanted. Their expectation that each ironmaker would keep his product unchanged discouraged innovation at the ironworks. Until metallurgists demonstrated how iron made to speci-

Fig. 8.5. Hephzivalla, the home of company vice president C. W. Barnum in Lime Rock. (Connecticut Magazine, vol. 8.)

fied composition and cast according to defined conditions would yield specified results, the Salisbury makers could continue to charge premium prices for their product. Slow metallurgical progress and founders' fear of tinkering with technique that worked kept a market for Salisbury iron going into the twentieth century.

Both professional visitors, such as Benjamin Silliman and Charles Shepard, and the Salisbury mine proprietors thought of the district's ore resources as inexhaustible. A. L. Holley remarked in 1877 that although production at Ore Hill had continued for many years, the ore was "abundant and apparently inexhaustible." However, miners had to dig ever deeper to get at the ore (fig. 8.6). By 1899 they were working at depths of 200 feet in galleries at Ore Hill. They hauled ore to the surface on an incline at the west side of the pit, where they crushed and washed it with water delivered by the pump that drained the mine. Companies of miners who paid royalties to the owners did the mining until 1898, when Barnum-Richardson took over operation of the Ore Hill mine.[17] John Maloney, who lived at Ore Hill, described mining as he remembered it in the 1890s. Until 1885 all the work was in open pits. Because of the dip of the ore beds, removal of the overburden became more expensive as the miners dug deeper. About 200 men worked underground, all digging by hand to expose the ore on a face about sixty-five feet thick and about three-quarters of a mile long. The thickness diminished with depth. A slope ran down the dip of the ore for about 430 feet. Miners drove tunnels north and

Fig. 8.6. The Connecticut Western Railroad crossed the Ore Hill mine pit on a trestle. Steam-powered pumps kept the pit sufficiently free of water for the miners to work. (Photo from the C. R. Harte collection, courtesy of Fred Chesson.)

south from the slope across the face of the ore and opened drifts to the east and west. They left pillars about thirty feet square to support the roof and later robbed the pillars by blasting.

A crew of four men (two miners and two wheelers) worked in each drift. They pushed the ore out of their drift in steel cars on rails that led to the slope. A skip on a cable went to the surface and carried the ore to the crushing and washing plant. Railcars took the ore to East Canaan; teamsters hauled it in horse-drawn wagons to Lime Rock and other furnaces not on a rail line. The drifts were damp, and the men pale for want of sunlight. Many died before the age of sixty. There was much negligence by the Barnum-Richardson Company, as well as carelessness among the miners. Few used strippers to fasten caps to fuses; instead, they clamped them on with their teeth. They passed dynamite sticks man to man. The timbering was sometimes bad. Yet there were few accidents. Heavy rains in the spring of 1897 flooded the mine as water in the north (Mammoth) pit spilled into the main pit. Men had trouble getting out as the water came in. In 1903 or 1904, two men were trapped by a fall of earth, and only one was rescued. In 1900 the company paid wages of $1.35 per day, six days a week. Although the miners had no union, they nevertheless staged occasional strikes.

The village at Ore Hill had a post office, railway depot, store, and school. The Barnum-Richardson Company built a clubhouse for the miners and their families in 1909. About 80 percent of the miners were Irish, and the rest, mostly Cornish. Everyone had gardens, chickens, cows, and swine that got them through slack times with adequate food.[18] By the summer of 1898, about thirty men working in underground galleries 150 to 200 feet below the surface could supply the furnaces in East Canaan with eighty tons per day of washed ore.[19] Nevertheless, as miners worked to greater depths, costs at Ore Hill increased. Consequently, in 1912 Barnum-Richardson reopened the old Weed mine across the border in New York, leaving Ore Hill nearly shut down.

From 1872, when the Roxbury furnace was blown out, until the depression of 1892, Salisbury ironmakers closed only one blast furnace, the Forbes, built in 1832 and rendered surplus by construction of the Canaan No. 3 furnace. Depressed iron prices in the 1890s forced owners to close the less-profitable furnaces. They blew out the Kent furnace in 1892, bringing mining in Kent, which had begun before 1732, to an end. When the blowing engine at the Chapinville furnace failed in 1897, the owners decided not to make repairs and instead abandoned the plant. Furnaces that lacked rail connections for delivery of ore and fuel were vulnerable. Buena Vista closed in 1893, Cornwall Bridge in 1897, Sharon Valley in 1898, and Lime Rock in 1900. Each furnace closure threw men out of work and rippled through the industrial and agricultural economy. The Cornwall Bridge furnace, for example, employed a hundred men and a hundred animals when it was making 3,000 tons of iron per year.[20] As the owners abandoned these furnaces, they let all their woodlands revert to the towns for taxes.

The all-steel car that railroads started buying in 1896 posed a more serious long-term threat to Salisbury ironmaking than depression. Because of the increased carrying capacity of these cars, the railroads needed stronger wheels. As soon as Pennsylvania steelmakers began forging and rolling solid-steel wheels in 1903, railroads began phasing out the cast-iron wheels that made up the principal demand for Salisbury iron.[21] This, along with metallurgical knowledge that allowed founders to use iron made with mineral fuel in place of the costly charcoal-made iron, meant a declining market for the Barnum-Richardson Company's products.

By the turn of the century, Barnum-Richardson got about 10 percent of its charcoal fuel by rail from Vermont and depended on local sources for the rest. All this charcoal was made in pits or kilns that recovered no by-products from the distillation process. By 1880 charcoal ironmakers elsewhere had adopted retorts and found that the by-products of wood distillation contributed substantially to their incomes. Finally, in 1914 Barnum-Richardson organized the Connecticut Chemical Company to make charcoal at a wood distillation plant in East Canaan. It had the capacity to process 21,000 cords per year. From one cord of seasoned wood the chemical company got 45 bushels of charcoal, 230 pounds of acetate of lime, and 11.5 gallons of alcohol. It processed 21,000 cords of wood a year delivered by up to 25 railcars a day, arriving from neighboring New York and Massachusetts, as well as Vermont and the company's 9,000 acres of Connecticut woodland.[22]

In 1915 failure to control costs and keep up with developments in mining and smelting technique brought the Barnum-Richardson Company to near bankruptcy. The non-resident owner, W. M. Barnum of New York, then "got mad" at his managers at the Lime Rock office and brought in a professional engineering consultant, J. E. Johnson Jr., who found that the basic problem was a plant more than twenty years out of date.[23] The company survived on the reputation of Salisbury iron, which metallurgical science was rapidly reducing. At the Ore Hill mine, pillars left in the past hindered efficient working. Miners still did their drilling by hand rather than with the air-powered drills commonly used elsewhere. A high phosphorus content in the ore was a problem. Although the company's woodlands were scattered, they were adequate for the fuel supply. Johnson felt that although the charcoal-producing retort plant was good, the by-product distillation plant needed improvement and an experienced manager. His report noted the need to season wood before retorting.[24]

Johnson found many deficiencies in the company's blast furnace operation, which was at least thirty years out of date. The newest furnace (Canaan No. 3) still had an open top and was filled from hand barrows. The fillers could not distribute the stock symmetrically in the shaft. The furnace flame was drawn into the hot-blast stove so that the air temperature could not be controlled. Most of the furnace gas was wasted instead of being used to fire the boilers of the auxiliary blowing engine. The keeper had to stop the blast every time slag was tapped. Johnson recommended that bell valves be installed and that the ore should be graded in a laboratory and stocked so as to keep different grades separate—all procedures adopted much earlier by operators of charcoal-fired furnaces in Alabama and Michigan. He believed that the fuel ratio could be reduced from 1.0 to the 0.82 or better achieved at other furnaces. The foundries at Lime Rock had no modern equipment. Haulage over the 2.5 miles between Lime Rock and the closest railroad station was a large expense.

Johnson's report was a litany of the faults of a company in decay. The Barnum-Richardson organization was top-heavy, with an extravagant proportion of earnings going to unnecessary salaries. When the firm had been prosperous, no one had made any effort to control costs. The staff were isolated and resistant to change. Johnson felt that one or two technically trained men should be hired to run the manufacturing end of the business. Resident manager R. N. Barnum's long personal acquaintance with the staff hindered his efforts to make needed changes. Former employee Charles Perkins later recalled how Bruce Hubbard in East Canaan was among the staff who deceived Barnum: Hubbard sent poor wood to the distillation plant and sold the good in New Haven "for his own rakeoff."[25]

High iron prices during World War I allowed Barnum-Richardson to begin a belated attempt to modernize, starting to build a new blast furnace (still in traditional style, however) in June 1918. Then, when iron prices fell in 1919, Barnum-Richardson plunged into receivership, and had to abandon new construction and close the Beckley furnace. Salvagers carried away the brick casting house and other materials.[26] In a January 1920 reorganization, creditors accepted shares of preferred stock in the Salisbury Iron Corporation in full settlement of their claims. They became owners of about 9,000 acres of woodland, three mines, three ancient blast furnaces, a foundry, and ninety dwellings for housing employees. The new owners included many Salisbury district families, among them Barnums, Richardsons, and Scovilles, that had prospered with ironmaking.[27]

As Johnson's report explained, there was little basis to suppose that Salisbury iron could continue to command the premium price that alone could justify the large expense of making it. Nevertheless, the Salisbury Iron Corporation managed to turn a profit on its foundry operation in its first year. However, it lost money on pig-iron production, primarily because of high costs of mining at Ore Hill. Steam engines with coal-fired boilers pumped water from the mine and hoisted the ore. The company cut its power cost from $2,250 to $1,600 per month by installing electric pumps. It investigated the possible sale of slack ore that had accumulated at the mine for over a hundred years. Engineers thought that with wet screening this material would contain 40 to 50 percent iron. However, no one could find a buyer for the slack. The company attempted to solve its fundamental problem, justifying the high price of its iron, by commissioning R. Moldenke, a well-known specialist in cast-iron metallurgy, to write a booklet on the virtues of charcoal iron.[28] None of these measures proved adequate, and by 1923 the corporation could not pay its debts. The Salisbury Iron Corporation then closed the Canaan No. 3 blast furnace. Inept management and the high cost of using renewable resources, which were still available, brought ironmaking in the district to its end.

NINE

A Landscape Transformed

Fluctuations in the national economy buffeted the Salisbury iron industry, but choices Salisbury's own ironmakers made about metallurgical technique determined both its course and its demands on the environment. A year-by-year count of the number of Salisbury forges and furnaces shows the rise and decline of the district's ironmaking, modulated by fluctuations in the national (or, earlier, colonial) economy.[1] The district's bloomery forges (fig. 9.1) made the wrought-iron products most wanted in the early eighteenth century. Because of the limited demand for castings (as well as the large investment required), a single blast furnace sufficed in the district until 1810 (fig. 9.2). By then, the Salisbury ironmakers had entered the market for high-quality wrought iron made by the indirect process and needed to enlarge the supply of pig iron for the new finery forges that began to supplant the old bloomeries (fig. 9.3). Local entrepreneurs added two new furnaces. By 1848, sixteen furnaces met the demand for pig from the additional finery forges built from 1825 through 1833, together with the requirements of the new puddling works. The smaller furnaces that specialized in making forge pig lost their market as the fineries, followed by the puddling works, closed in the 1850s. The remaining furnaces, making pig iron for foundries that specialized in chilled-iron railroad wheels, carried on until the railroads' adoption of steel wheels curtailed this market in the twentieth century.

The national ebb and flow of business, along with disruptions caused by war, modulated the trends established by the techniques the Salisbury ironmasters chose and the types of products they sold. Investment in bloomeries accelerated during the colonial prosperity of the 1740s and slowed during the wars with the French and the Revolution. Return of settled times in the early Republic led many individuals and partnerships to build bloomery forges in the years up to 1807 and to invest in furnaces and finery forges. Hard times after the War of 1812 suspended new investment. The period of the district's greatest growth fell in the economic expansion from 1824 through 1837, when New England entrepreneurs made rapid progress in developing the American system of manufactures based on interchangeable parts and power-driven machine tools. The panic of 1837 arrested growth, but most Salisbury works survived to thrive again in

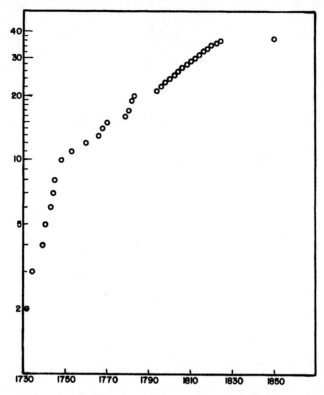

Fig. 9.1. *Cumulative number of bloomery forges built in the Salisbury district. Because the exact construction dates for nearly all the bloomery forges built after 1790 are not known, they are shown distributed year by year. Few closure dates are known; the number remaining in use probably declined rapidly after 1830. (Based on data in table A1.2).*

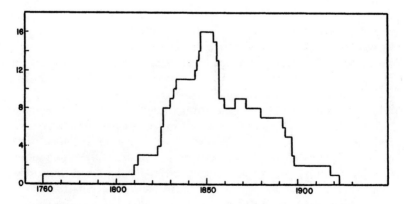

Fig. 9.2. *Number of blast furnaces in operation in the Salisbury district. (Based on data in table A1.1.)*

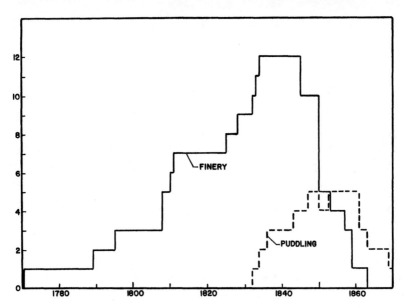

Fig. 9.3. Number of finery forges (solid line) and puddling works (dashed line) in operation in the Salisbury district. (Based on data in Tables A1.3 and A1.4.)

the 1840s. However, the panic of 1857 hit the Salisbury industry hard because the forge proprietors had allowed their works to drift into obsolescence. The Civil War did little to revive the industry. Pig iron production continued at the remaining blast furnaces until depression hit in the 1890s. By 1900 only two furnaces remained in use.

The basic problem that Connecticut ironmasters faced since the early days of the Republic, when they began selling iron nationally, was that ironmakers in other regions with better factor advantages could capture Salisbury's customers unless the Connecticut makers offered a product sufficiently better to justify a higher price. To succeed, a forge or furnace proprietor had to make the best possible use of artisans' craft skills and knowledge. Samuel Forbes, the first to build Salisbury's reputation, had been both a skilled artisan and an astute manager. However, he never had to adopt new techniques: well-established methods of ironmaking in use since the end of the sixteenth century sufficed. Once forge or furnace proprietors elsewhere began using new methods and applying developments in metallurgical science to improve their processes or products, Salisbury ironmakers faced new, unwelcome challenges. However, for a while they could shelter under the supposedly special properties of Salisbury ore.

In his 1856 essay on the Ames Iron Works, Zerah Colburn touted the superiority of wrought iron made from Salisbury pig, reiterating the belief held by district ironmakers that particular ores imparted particular properties to fined or puddled metal.[2] Forge managers believed gun iron had to be made from Old Hill ore alone. We now know that the quality of the

Salisbury products that built the district's reputation reflected artisans' skills in smelting and fining rather than any special property of the ore.[3] Nevertheless, the reputation of Salisbury iron, whatever its real source, served the district well for over a century.

Even with their industry in decline, people in the district still wanted to believe in the unique value of Salisbury ore. They thought that if they owned mine property they could collect royalties without having to bother with the business of making iron. Ore dug from the Davis mine in 1855 sold for $2.50 to $2.75 per ton. Miners got half of this money, and the other half went to the mine proprietors, organized as the Davis Ore Bed Company. In 1855 the company had receipts of $4,870 and, after paying its expenses for opening a drain and surveying, paid out dividends of $4,050 to proprietors, who included Lee Canfield and the heirs of J. C. Coffing and J. M. Holley.[4] These people collected their money without undertaking managerial responsibilities. The successors to the original proprietors still took their royalties from Ore Hill in 1872, receiving 75 to 100 percent dividends on the original investment made more than half a century earlier. An association of proprietors owned the land, and a mining company took out the ore and paid a royalty until 1898, when the Barnum-Richardson Company bought the shares of both.

The profits that mine proprietors made stimulated continued exploration during the prosperity of the early 1870s. George H. Holley used a steam drill to explore down seventy feet on lands belonging to S. S. Robbins and the heirs of A. H. Holley.[5] George Peet promoted opening the Bushnell ore bed, located two miles north of the Davis bed, in 1871. In one of the few examples of the use of outside experts, he called in Professor Verrill of Yale to evaluate the Bushnell mine. Peet was a bank president and probably knew little of the iron business. When he got a report he did not like, he denigrated Verrill's advice, claiming it to be of less value than the opinions of old-time miners.[6] In Norfolk, adventurers were sinking a shaft at Crissey's old iron mine.[7] None of these prospects proved of much value, leaving the three established mines to continue to supply most of the region's ore.

By 1870 all the district's ironworks owners had abandoned their finery and puddling forges instead of adopting the new methods of making wrought iron, such as the charcoal-hearth process developed by Pennsylvanians. They did this at a time when Americans were increasing their consumption of wrought iron, as they would continue to do over the next thirty years, since engineers continued to prefer charcoal-hearth iron to steel for some critical applications on into the twentieth century. The Salisbury forge managers had dropped out of a growing rather than contracting market for wrought iron. An often-repeated legend holds that America's adoption of Bessemer steelmaking (introduced, ironically, by A. H. Holley's son A. L. Holley in 1866) led to the demise of the Salisbury district's wrought-iron business. Actually, Salisbury makers had never competed in the market for tonnage production of iron that Bessemer steel captured. Instead, they had concentrated on the high end of the market, which steel took over only decades later.

Had they wanted to continue making a different high-value, low-volume product, the Salisbury ironmasters could have entered the shear and crucible steel business, as the nearby Collins Company did. Transportation costs were a minor consideration in crucible steelmaking, while artisans' skills and meticulous attention to quality were essential. It was the sort of business that Salisbury had excelled in. An entrepreneur named Broadmeadow scouted the area for a steelworks as early as 1839 but met with no response from the district's entrepreneurs.[8] Instead, Salisbury ironmakers sheltered behind the ancient reputation of the district's ore.

With closure of the forge branch of ironmaking, people began leaving the Salisbury area, beginning a decline that would reduce the number of residents in the district by a third by the time the last furnace closed (table 9.1). The people who stayed on in the district began transforming its land to new uses. They had long been aware of the physical attractions of their region: John M. Holley wrote home about the beauties of Salisbury while visiting Harpers Ferry, and his son, Alexander H. Holley, described them in his diary (see chap. 6). The decision of the residents of Furnace Village in 1846 to change the name of their town to Lakeville and again call Furnace Pond by its Indian name, Wononskopomuc Lake, indicated their changing attitude toward ironmaking. By then city dwellers considered Salisbury a romantic place to visit. In 1854 *Gleason's Pictorial Drawing-Room Companion* described the beauties of the scenery along with the sights to be seen at the Chapinville blast furnace and the Ames ironworks. Its writer placed the Ames works among "the objects of curiosity . . . apt to tempt you back again and again." The author was particularly impressed by the great steam hammers at Amesville (see fig. 7.8).[9]

Table 9.1. Population Data

Year	Population
1756	4,139
1800	11,613
1810	11,962
1820	12,640
1830	12,969
1840	11,990
1850	13,772
1860	13,641
1870	13,853
1880	13,612
1920	9,725
1930	10,559
1990	17,751

Note: These census data are for the towns of Canaan, Cornwall, Kent, Norfolk, North Canaan, Salisbury, and Sharon.

Henry Ward Beecher, pastor of the Plymouth Congregational Church in Brooklyn, New York, described the Salisbury landscape where he spent his 1853 vacation in a series of letters known as the Star papers. He arrived after a five-hour ride on the Harlem Railroad, which had reached Millerton two years earlier. Beecher was impressed by his landlord's refusal to sell off his forty acres of woods despite insistent demands from developers. The landlord valued the woodland highly. In addition to fishing for trout in most of Salisbury's streams, Beecher explored the hills by rides along colliers' roads. He wrote enthusiastically of Sage's Ravine, the Great Falls, and the Twin Lakes. At this time the Great Falls had the Ames ironworks with its two great steam hammers, the Canfield & Robbins forge, and the Canfield blast furnace in full operation. The Chapinville blast furnace smelted iron at the outlet of the Twin Lakes. Woodsmen regularly cut over the hillsides throughout the region for fuel wood (fig. 9.4). Beecher gives no hint that these industries interfered with his enjoyment of the region's natural beauties.[10]

While people in Salisbury turned more and more to the accommodation of visitors after mid-century, the ironmakers continued to make heavy demands on the region's forest and ore resources. In 1817 Benjamin Silliman had noted immense forest, "the consumption of which would seem to be beyond the power of any population which is ever likely to accumulate in these regions."[11] Nevertheless, ironmakers had already committed most of the district's woodland to growing fuel by 1830. Although the railways could deliver mineral coal from Pennsylvania to the Salisbury district, their first effect on the fuel supply was increased demand for fuel wood: New England railroaders liked wood-fired locomotives and resisted

Fig. 9.4. Woodsmen have cut coppice wood from this Cornwall hillside and stacked it for colliers to convert to charcoal. (Courtesy of the Cornwall Historical Society.)

changing to coal. Additionally, sparks from locomotives often started fires in woods along the tracks, placing further pressure on the wood supply.[12]

Increased thermal efficiency in the region's furnaces could reduce fuel consumption. District ironmasters had already achieved their largest, and easiest, improvement in average fuel ratio at their blast furnaces, from 1.4 in 1837 to 1.0 in 1860, by adopting hot-blast stoves and retiring old furnaces. They were still operating at a fuel ratio of 1.0 in 1916. At this consumption rate, the district's furnaces used the wood from 3.87 square miles of forest in 1860 and from 6.1 square miles in 1870.[13] Completion of the Connecticut Western Railroad in 1871 allowed Barnum-Richardson to import about 10 percent of its charcoal from Vermont. By then the company had lost the option of substituting mineral coal for charcoal, since it survived on the premium price it charged for charcoal-smelted iron. It needed about 100 square miles of woodland in continuous production to sustain its furnaces. Another 10 to 15 square miles were needed to provide the fuel wood most people still used to heat their homes. Locomotive fuel would have required further area until the railroads adopted coal fuel. To meet these demands, over a third of the 317 square miles of the seven towns that supplied most of the charcoal for ironmaking would have been committed to growing fuel wood. However, this was not the burden it might have been earlier because farmers were letting former pastureland return to forest (table 9.2).

Miners had to dig deeper pits to get the ore to sustain pig iron production after 1860. Charles F. Perkins went to work in the Davis pit in 1892. Here the ore was in wedges dipping at about forty-five degrees, which the miners reached by a slope. They used a skip hoist to haul ore and overburden out of the pit. The ore went to a washer, and the overburden went to dumps around the pit. Numerous prospecting holes pockmarked the mine area, and mine waste formed surrounding hills.[14] Julia Pettee, Salisbury native and descendant of the Mount Riga ironmaster, described the Davis mine just before Barnum-Richardson closed it in 1904 (fig. 9.5). The dumps of overburden formed large mounds of yellow earth with roads spiraling to their tops. Two-wheel horse-drawn carts carried out the waste. A steam engine housed above the pit on the east side pulled skip cars loaded with ore from the pit to the crusher and washer. The stream

Table 9.2. Percentage of Forest Area in the Town of Norfolk

Year	%
1818	64
1850	41
1860	44
1870	50
1914	77
1931	84

Source: Herbert H. Winer, "History of the Great Mountain Forest, Litchfield County, Connecticut," Ph.D. diss., Yale University, 1955, p. 159.

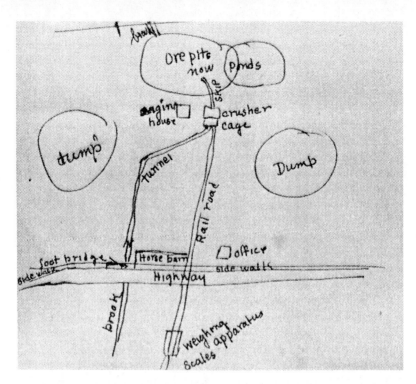

Fig. 9.5. Map of the Davis mine in Lakeville by Julia Pettee. Spoil piles surround the mine pit. The brook passing under the road carried away waste from the ore washer. (Manuscript, Scoville Memorial Library, Salisbury.)

that flowed through the mine provided water to wash the ore, and a tunnel lined with planks carried the waste under the highway. The stream of turbid, yellow water that emerged deposited silt along the tracks that led to the mine and the adjacent meadow.

By 1960, fifty-six years after mining ended, A. J. Rosenthal had developed an estate on the east side of the Davis mine waste dumps, and the town had built its high school adjacent to those on the west.[15] Today the water flowing under the road is clear, and it is hard to find any trace of the silt deposits.

Slag was the principal waste made at the blast furnaces. By the 1870s, no finery forges remained to recycle the shot iron in the slag, and instead furnace operators dumped it in piles on the nearest convenient waste land. In East Canaan this was across the Blackberry River for the Forbes and Beckley furnaces. In 1915 the New England Slag Company set up a plant to recycle this waste by crushing it into sizes useful for aggregate and for coating fireproof shingles. It recovered the shot iron that had been entrained in the slag. For a while the slag company shipped out a dozen or so carloads a day. It processed all the slag left by the Forbes furnace and about half that from Canaan No. 3 before going out of business.[16] Town highway crews still dig road fill from the slag banks from time to time. The still-substantial remaining slag piles from the Beckley furnace can be found in the woods today with a bit of searching.

While ironmaking continued its demands on the environment through the 1870s and 1880s, Salisbury people accelerated the transformation of their land to the new uses that had been initiated in mid-century. As they did this, visitors began to see the mines and furnaces as quaint features of the landscape. Thus, in 1874, the editors of *Picturesque America* represented the Kent furnace as a romantic, somewhat decrepit survivor from an earlier time (fig. 9.6; cf. fig. 8.4). Individuals who had left Salisbury for jobs elsewhere began to return to the countryside they loved.

Fig. 9.6. *Romanticized view of the Kent blast furnace. When this drawing was made in 1874, the furnace would have looked like the Buena Vista furnace shown in figure 8.4. (From* Picturesque America, *vol. 2, New York: Appleton, 1874, p. 291.)*

George W. Holley, who left his position as a partner in the Lime Rock branch of the family iron business in 1837 for western New York, came back for fly-fishing and walks through the region's scenery.[17] Alexander Lyman Holley, the designer of steelworks throughout the United States, had hoped to build a home in Lakeville because of his love of the landscape. (His early death thwarted this plan.)[18] Jonathan and Nathaniel Scoville grew up in Chapinville in the 1830s, when a forge and blast furnace were at work, ran the furnace in Huntsville in the 1850s, and made a lot of money with a foundry and other enterprises in Buffalo, New York, in the 1870s. A decade later they returned to Chapinville, where they bought a thousand acres for a model stock farm adjacent to the still-operating blast furnace. The Scovilles also continued the local penchant for speculating in iron mines: they thought that ore from a bed on the nearby mountainside was particularly suitable for stove castings. They evidently saw no conflict between their estate and a mining operation nearby. In 1881 they surveyed a route for a railroad spur and put some miners to work, but they soon realized the futility of this project and devoted themselves to developing the agricultural side of their estate. They took over the debts of Landon & Company, proprietors of the blast furnace, razed the furnace, and used its water privilege for a small hydroelectric station that supplied their mansion with electric light. Although the original house of the Scoville estate burned in 1917, its replacement remains a landmark in the region's transformation today.

Salisbury was south of the Berkshire region of Massachusetts, where wealthy families from New York and other eastern cities began building estates in the 1840s. Wealth acquired in the Gilded Age allowed many owners to enlarge their properties for ostentatious display. Some of this estate building penetrated into the Salisbury district, principally in Norfolk. Completion of the Shepaug Railroad encouraged city residents to build estates in Washington. The railroad changed some established place names to honor its wealthy, new customers, as when it renamed the Davies Hollow stop "Romford" in honor of Romulous Ford, one of the new landowners.[19] Over in Lime Rock, the Barnums and the Richardsons built their smaller but still impressive mansions with the money they made out of Salisbury iron (see chap. 8). Litchfield saw less of the Gilded Age mansion development, partly because its surviving eighteenth- and early nineteenth-century dwellings were already sufficiently elegant to deter the newly rich from tearing them down. Instead of building mansions, some of the new Salisbury district residents bought up large acreages of forestland for recreational use and conservation as the ironmakers abandoned their woodland to the towns for unpaid taxes. A fishing club bought the forest owned by the Hunt-Lyman Iron Company in Canaan and Norfolk to hold as a preserve. The Childs-Walcott families acquired 6,550 acres of former charcoal-producing land known today as the Great Mountain Forest.[20] The last large block of iron company land, the Shepaug Iron Company's 400 acres, remained in private hands until acquired by the Roxbury Land Trust in the 1980s. For these various reasons, the Salisbury district escaped the grosser manifestations of the wealth brought into northwestern Connecticut and neighboring western Massachusetts by newly arrived people from the eastern cities.[21]

Private schools, with their expensive buildings and large, landscaped acreages, made an enduring contribution to transformation of the land. Hotchkiss in Lakeville, founded in 1892, was followed by the Salisbury School (1901), the Connecticut Junior Republic (1904), Kent (1906), Rumsey Hall (which moved to Cornwall in 1907), Canterbury (in New Milford, 1915), and South Kent (1923).[22] Cultural institutions such as the Yale summer music school at the Stoeckel estate in Norfolk and Music Mountain in Canaan, devoted to performances by a resident string quartet, joined the schools in transforming more acreage to new uses. The flow of writers, artists, and musicians to the northwest that started in the 1920s swelled in the 1930s.

Lime Rock's fortunes had been closely tied to the Barnum-Richardson Company. Its demise left the village virtually deserted after the foundry closed in 1923. Three of the ornate Barnum and Richardson houses soon burned to the ground. Alfred Stone of New York City purchased the 326 acres, twenty-seven remaining houses, and assorted outbuildings owned by the bankrupt Salisbury Iron Corporation in 1825 and then sold many of the properties to artists and writers. One of these participants in the Arts and Crafts movement converted the foundry to a paper mill, another made prints on linen in the Casino (the recreation building Barnum-Richardson put up for its employees), and a third made etched books. Alfred Korzybski set up his Institute of General Semantics in the former Leonard Richardson mansion.[23]

The prolonged transition from primary industry to residential and institutional use of the land gave Salisbury residents time to adjust to new economic opportunities. Some members of the ironmaking community made this transformation more slowly than others: the colliers and woodcutters who worked in the late nineteenth and early twentieth centuries, mostly recent immigrants from Canada and eastern Europe, lacked the social skills, education, and experience of earlier colliers, and many retreated to isolated lives in the hills. As early as 1917, writers began to romanticize the people who lived in old iron- and charcoal-making communities in the mountains, such as the remaining population on Mount Riga, who appear as uncouth hill folk in Edith Wharton's 1917 novel, *Summer*. By the 1930s the surviving families in the backwoods were known locally as "Raggies." In 1938 the essayists I. H. Sterry and W. H. Garrigus portrayed them as living in squalor, drinking nothing but cider brandy while eating woodchucks, stolen chickens, and poached deer. They described the Raggies as a proud people, whose women had a child each year throughout their reproductive lives.[24] In 1946 C. P. Smith described

Fig. 9.7. *Photograph on Mount Riga looking north taken about 1870, twenty-five years after closure of the furnace and forges. At this time the Barnum-Richardson Company owned the land for coaling and used the pond as a reservoir of water for its blast furnace at Lime Rock. Stonework of the South Pond dam and Woodin's forge are at the left; the ironmaster's house is across the hillside. (Courtesy of the Salisbury Association.)*

Fig. 9.8. *Modern view of the scene shown in figure 9.7. The light area at the far end of the dam is an artificial beach.*

the Raggies on Mount Riga as "inbreeding as rapidly as possible." He saw them feeding themselves by fishing and hunting, and as skilled wood-workers who would steal liquor and small sums of money wherever they worked but were never responsible for any greater crimes.[25]

By 1950 a visitor would have been hard-pressed to find examples of Rag-gies if, in fact, they ever existed outside the imaginations of novel writers. The transformation of the land was essentially complete. Mount Riga epitomizes the change. A picture taken about twenty-five years after the furnace and forges closed shows land still in coppice wood production (fig. 9.7). The stone-faced South Pond dam stood next to the foundations of Woodin's long-abandoned forge. Recently cut woodland and rough pasture surrounded the ironmaster's house. City visitors began buying up Mount Riga for summer cottages early in the twentieth century and even-tually organized the Mount Riga Corporation to manage the 4,200 acres of former iron-company land they acquired on the mountain. Today the cutover land is reforested (fig. 9.8). The dam at South Pond is regraded and grassed, with a bathing beach at its end. Roads, although still dirt-surfaced, are widened and smoothed. The ironmaster's house preserves traditional Salisbury values as a summer residence (fig. 9.9): it is essen-tially unaltered; not even plumbing has been installed. In the woods down

Fig. 9.9. The ironmaster's house of Mount Riga stands today unaltered except for maintenance of the original structure.

Fig. 9.10. This massive dam formed the pond for the lower Mount Riga forge, which stood to the left of the stream. A flume fitted into the notch on the crest of the left side of the dam carried water to the forge wheels.

the river stand the remains of the great lower forge (fig. 9.10) where artisans once made anchors for frigates and the gun iron the national armories converted into the muskets and rifles carried by American soldiers
in three of the nation's wars.

Community, Culture, and
Industrial Ecology

The people who settled northwestern Connecticut created an agricultural surplus that allowed them to undertake industrial ventures within a few years of their arrival. Their knowledge of the mechanical arts, coupled with the region's natural resources, gave them opportunities to make material goods needed by their neighbors. Successive generations continued industrial use of the region's natural resources over the next two centuries, each making its own choices about how to structure its enterprise within the framework of values and beliefs held separately by individuals and in common within the community. Each had to respond to changes in markets and the advent of new products and techniques. These opportunities, and the participants' choices about how to use them, combined to create the region's industrial ecology.

Like the rest of the New England hill country, northwestern Connecticut had two abundant, renewable natural resources: streams with steep gradients and reliable flow for waterpower, and forest that covered the large areas that were too steep or too thinly mantled with soil for decent pasture. Millwrights could easily build waterpower systems on the streams, and farmers could manage the forest for continuous production of fuel wood, since it regrew trees to useful size within about twenty years. Unlike other highlands, however, northwestern Connecticut had a unique mineral resource: iron ore beds unmatched elsewhere in New England.

Everyone in the newly settled lands and on the frontiers expanding into Vermont and New York in the early eighteenth century needed iron products. As described in chapter 3, individuals throughout the Salisbury district, aided by family members or fluid partnerships, built bloomery forges that they operated as components of their cropping, husbandry, or mercantile enterprises. Nearly every family in Kent and the other new towns had a partner in one of the forges. Individuals lacking metallurgical skills or access to any capital dug ore or cut wood. Others developed their skills as colliers or millwrights. Negotiated exchanges of labor and services among these artisans promoted interdependence within the community.

As the colonists in southern New England increasingly mechanized their grain, timber, and cloth production in the mid–eighteenth century, they brought a new opportunity to the ironmakers of the Salisbury district. By making standard parts for grain mills, sawmills, fulling mills, and oil mills that they could distribute widely, Salisbury ironmakers added value to the bar iron they made and enlarged the scope of their market. As they chose to specialize in these high-value-added products, they took the first step toward differentiating their district from other ironmaking areas in North America.

Investors with profits from land sales gave people in the northwest a further opportunity to redirect the course of their ironmaking in the 1760s. The Lakeville blast furnace made the first change in the region's dispersed, small-unit mode of production. Samuel Forbes of Canaan solved the technical problems of building and blowing in the furnace, but it took an outsider with wide trading experience, Richard Smith, to make a financial success of the furnace project. Once Smith had shown the way, Salisbury's own entrepreneurs, working in partnerships like those they had used for their earlier bloomery forges, built additional furnaces.

As they shaped their industry, the proprietors of Salisbury's forges and the furnace turned to more sophisticated management of the region's natural resources. They began acquiring large acreages of woodland they could manage for continuous production of wood fuel and began incorporating the region's lakes into their waterpower systems to gain a reliable supply of mechanical power for their works. They worked out cooperative agreements to permit multiple use of the water released from their reservoirs as it passed successive downstream mill sites.

When Americans undertook mechanized manufacturing of wood and metal products, and increased the size and scope of their industries during the opening decades of the new Republic, they gave Salisbury ironmakers more opportunities to participate in the nation's growing economy. They could respond to increased demand both for primary materials and for more sophisticated, higher quality products. Since they had access to deepwater ports on the Hudson River, they could sell iron products on the national market in competition with ironmakers in the Middle Atlantic states, if they wanted to. John M. Holley and his partners made initial moves in this direction by shipped pig iron to Boston and iron water-main pipes to Albany. However, by 1808 they had already concentrated on smelting pig for Salisbury's new finery forges built by the Rockwells, Boyds, and Cooks, joined later by Holley & Coffing. These forge proprietors defined the direction the district's iron industry would take through mid-century: they opted for the premium-price, small-volume end of the market, where they could sustain production with the region's renewable energy resources. We would like to know what factors led the Salisbury makers to this choice.

Resident artisan-entrepreneurs and merchant capitalists with strong community ties made the decisions about the course of ironmaking in Salisbury through its first century. The region had no absentee ironworks owners because, in the eighteenth century, the Forbes, Adam, Holley, and other Salisbury families bought up the interests of the out-of-district investors whose capital had helped open the mines and start smelting.

These Salisbury families earned enough from their farming, trading ventures, real estate dealings, and ironmaking to build their industry on their own. Additionally, individual artisans who accumulated capital from farming, smithing, or mining could undertake iron smelting. Thus, the Hunts pooled their family resources to start the forge business noted by Benjamin Silliman on his 1817 visit, a business that would endure for two-thirds of a century. Other families bought shares in the large ore beds, thereby establishing the basis for the prosperity of several future generations. Local ownership, often coupled with family connections, dominated in Salisbury, while elsewhere (eastern Pennsylvania, for example) non-resident investors often gained control of ironmaking.

The Salisbury owner-artisans, such as the Hunts, Rockwells, and Boyds, typically worked at the forge or furnace alongside their finers and hammermen. Others, like the members of the Holley and Adam families, relied on artisans working as inside contractors who negotiated agreements for the work they would undertake. Salisbury artisans learned their art by experience. Most had a wide range of skills and could take on whatever tasks needed doing, unhindered by work rules or established rights, privileges, or traditions. Many artisans did carpentry, quarrying, milling, or storekeeping or worked as teamsters as opportunity offered. In a community where artisans might also be owners, and owners artisans, there were few social distinctions, and no large disparities of wealth. Social stratification remained relatively weak while the forge branch of the iron industry was ascendant. Thus, colliers, forgemen, and furnace hands often joined Alexander H. Holley at his house in the evenings to exchange stories and reminiscences. An artisan who hired helpers one day might work for others a few days later.

Everyone in the region, including lawyers and other professionals, farmed, often with considerable skill. Many ran gristmills and sawmills alongside their ironworks. Owners and artisans alike moved in and out of ironmaking as their preferences and opportunities changed. Everyone kept detailed records in account books so that they could settle up in cash any imbalance in the services they had rendered to each other. Lack of specialization freed the community from total dependence on the fortunes of one industry and gave residents alternative occupations to fall back on during slack times in the iron trade.

Liberal learning figured prominently in the Salisbury community. Since artisans as well as owners had to keep records, figure costs, and negotiate contracts, they needed basic education. From the seventeenth century onward, Connecticut statutes required parishes to maintain public schools. Salisbury parents often sent their sons and daughters to academies for further study, and many sons went to Yale or to colleges in Massachusetts. Malcolm Rudd found more teachers, writers, jurists, lawyers, and other professionals to include in his *Men of Worth of Salisbury Birth* than ironmakers. Young men formed literary associations, and wrote of the natural beauties of their region. The Congregational church remained an intellectual force (as well as a physical presence in the district's towns), strengthened after disestablishment in 1818 by the revival led by Lyman Beecher. New attitudes toward intoxicating beverages accompanied the growing intellectual life of the district. After 1820, the temperance movement made substantial progress throughout the community.

Despite their commitment to liberal education, Salisbury people rarely extended their reading to natural science or technology. Salisbury artisans made iron using seventeenth-century techniques, and their letters and business records hardly ever mention technical matters. Ironworks owners and managers had little use for the expertise of engineering professionals and rarely called one in. Alexander L. Holley's confrontation with his father over college shows the small place science and technology held in Salisbury's culture. Alexander H. Holley was unsympathetic to his son's interest in railroad technology and allowed him to study engineering at Brown rather than the classics at Yale only after a prolonged struggle. Salisbury furnace keepers, finers, and ironmaking managers knew little of the chemistry and metallurgy their counterparts in Pennsylvania were applying to increase productivity. They had built Salisbury's reputation with the care and skill with which they used established methods to make products of superior quality; they had little interest in new ironmaking techniques.

Concentration on quality rather than quantity placed a premium on artisans' skills and managers' close supervision. This choice made good use of the region's natural and human resources and, before construction of the Housatonic Railroad, lessened the difficulties that would have arisen in shipping large quantities of products overland during the spring, when the roads to Hartford, New Haven, or the Hudson River ports were bad. The choice fitted comfortably with the community and individual values that focused on the quality of life in Salisbury instead of prominence in national industry. Proof would be difficult, but it seems likely that these values entered the decision to eschew quantity production.

Horatio Ames's attempt to become one of the nation's largest suppliers of locomotive tires, the Canfield and Robbins scheme to bring a cluster of factories to Falls Village with their Water Power Company, and the Mine Hill steelmaking project organized by Hartford investors all had access to good natural and transportation resources. However, each aimed at a larger scale of enterprise than anything previously undertaken in Salisbury, and each fell outside the district's established way of doing business. The Mine Hill artisans, brought in from outside, never mastered the technological novelties the proprietors adopted for their blast furnaces and steelmaking. People in Salisbury had little enthusiasm for the community of operatives that would have been needed to staff the factories Canfield and Robbins visualized as clients of their Water Power Company. Ames's forge had the best prospects and prospered for a while, but it failed as Horatio overreached himself. Additionally, where J. M. Holley and other Salisbury forge proprietors had maintained respectful and civilized, if not always tranquil, relations with their government clients, Ames's intemperate and confrontational tactics antagonized the officials whose cooperation he needed.

Salisbury ironmasters could have continued working their finery forges for several decades after 1860, as some forge proprietors did in Pennsylvania, They could have gone into making crucible steel, a high-value, low-volume product still dependent on artisanal skill. They could have built modern charcoal-fired blast furnaces as Yankee entrepreneurs did in Alabama during Reconstruction in the South. They could even have used anthracite brought in by the Central New England Railway, which had a cross-Hudson bridge at Poughkeepsie, to fuel modernized blast furnaces like those at Port Henry and Standish in the New York Adirondack region. Instead, by 1870, the Holley, Boyd, Moore, Canfield, Rockwell, Robbins, Adam, and other ironmaking families that had established the district's reputation for the quality of its bar iron had dropped out of ironmaking. While some continued to collect handsome royalties from the mines, put money in model farms, or sell land to wealthy city dwellers for estates, few focused on ostentatious display of wealth. The one Salisbury native to remain prominent in ironmaking after 1865, Alexander Lyman Holley, worked from an office in New York City.

The Barnums and the Richardsons, who bought up the by-then-obsolete blast furnaces after mid-century, carried a new set of personal and community values, represented by the mansions they built for themselves in Lime Rock. In the 1880s their agents and artisans pushed the region's output tonnage to its maximum as they traded on Salisbury's by-then-ancient reputation to sell their iron at high prices. The dominant partner, W. H. Barnum, had already become an absentee owner deeply involved in politics and with no interest in metallurgy. He and his successors let the iron business drift into obsolescence and eventual bankruptcy.

Choices made in the context of Salisbury's particular culture controlled the region's industrial ecology. Salisbury ironmasters used renewable energy resources to process the non-renewable ore they mined. They managed forests for continuous fuel production and designed waterpower systems that continued to meet their energy needs. They always had plenty of ore.[1] The proprietors of the forges dropped out of business for reasons unrelated to their resource base well before they encountered limits, or excessive costs, arising from their choice of renewable energy. The Barnum-Richardson firm's subsequent exploitation of Salisbury iron's reputation did not test the limits of sustainable energy, and in the hands of competent management could have kept the northwest's forests in fuel and wood by-product production for another decade or more, as did the charcoal ironmakers in northern Michigan.

Although resource exhaustion was not an issue in the Salisbury district, use of natural resources for ironmaking brought environmental change on both short and long time scales. Owners, managers, and others making decisions about resource use could weigh the gains to themselves, their neighbors, and the community against obvious environmental costs as they directed the course of their industry. However, they left long-term consequences for future generations to deal with.

Woodcutting and coaling, mining, quarrying, hauling ore and charcoal, railway shipping, smelting, forging, and foundry work changed the environment in which people in Salisbury lived and worked. Benjamin Silliman noticed the great piles of rubbish stacked around the Lakeville furnace during his 1817 visit. Ironmakers put the region's forests into continuous fuel production, ore wagons and charcoal wains rumbled through the street of the district towns, and furnace operators piled up ever higher hills of slag. Yet Salisbury residents rarely (if ever) mentioned the environmental effects of ironmaking in their letters, newspapers, or reminiscences: they saw the working landscape as tolerable, or unremarkable.

Mid-nineteenth-century visitors to Salisbury commented on the spectacle of hot metal and heavy forge hammers rather than deforestation, smoke, dust, or piles of slag. These visitors came for fishing, mountain rambles, and peaceful relief from the hustle of the city. They found the forges and furnaces points of contrast adding variety to the region's scenery. When Henry Ward Beecher wanted to take a summer vacation in a region having mountains and fast-running streams well stocked with fish, he could as well have taken the Jersey Central Railroad to northeastern Pennsylvania as the Harlem Railroad to Salisbury. Many city people did travel to Pennsylvania to visit the spectacular scenery of the state's northeast corner and ride the great incline railroad at Mauch Chunk. Nevertheless, the owners of the Pennsylvania region's industry had already started along a route that would lead their region to a level of social and environmental degradation never experienced in Connecticut.[2]

In the Salisbury district, the long-term environmental consequences of nearly 200 years of mining and ironmaking—those we see today—are minimal. The deep mines are now lakes fringed by woods or lawns. New vegetation has transformed the smaller mines into wooded hollows in hillsides and has grown over the surrounding spoil piles. Silt deposited from the wash water that miners discharged onto surrounding lowlands is now wetland, grass covered, or has been built upon.

The heavy use of the region's renewable fuel and power resources has also left only small traces. Forest uncut since its last coppicing is now well wooded with substantial trees. One forge (Ames's) and three of the region's blast furnaces (Bradley's, Kent, and Bulls Bridge) used power from the Housatonic, where their dams ponded water within the river. Floods have carried away nearly all traces of these dams and races. Three other furnaces (Lakeville, Mount Riga, and Chapinville) drew water from existing lakes that now support recreational boating, fishing, and swimming. The region's other forges and furnaces drew on smaller streams, such as the Blackberry. Freshets have breached most of the forge and furnace dams and carried away any accumulated silt deposits. People use the surviving furnace ponds for still-water fishing, and reservoirs made to hold reserve water for furnaces now appear as much a part of the landscape as the regions's natural lakes.

Size, dispersion, recycling, and the chemistry of the materials used limited the environmental impact of the Salisbury iron industry. A region's principal topographic features serve as a scale against which to judge physical size. All of Salisbury's open-pit mines—the most conspicuous feature of the industry after the coppiced woodlands—were smaller than their adjacent hillsides. Those that are now water-filled are comparable to, or smaller than, nearby natural lakes. The hillside mine openings are less conspicuous than the gorges cut by the region's mountain streams. In contrast, the great iron-mine pits that dominate the landscape in parts of northern Michigan and Minnesota are larger than any neighboring topographic feature.

Size also limited the environmental impact of the Salisbury ironworks' waterpower systems. Each was small relative to an installation that would fully develop power that the Housatonic, the region's large river, could supply. Small streams that ironmasters could develop with modest capital provided adequate power for most of their forges and furnaces. The ironworks that used the Housatonic took only a fraction of the river's flow. People built hydropower installations with major environmental impacts only when outside, twentieth-century capitalists financed hydroelectric stations to generate electricity for export to cities elsewhere in the state. These power plant builders converted long reaches of the Housatonic into still-water basins and made the reservoir of the Rocky River pump-storage plant (in New Milford) Connecticut's largest lake.

Dispersion also limited short- and long-term environmental effects. The distances between ore beds were large compared with the size of the pits. Since no furnace or forge could be larger than a limit set by the power available from its water privilege, and the region had numerous but relatively small streams, its forges and furnaces had to be well separated. Aside from the harvesting of wood for coaling, the visible evidence of ironmaking remained in isolated spots spread over a large area.

Because Salisbury ironmakers recycled many of their wastes, they left relatively little debris behind. When they crushed blast furnace slag to recover the particles of iron it contained, they reduced the slag to sand-size debris that easily mingled with the region's glacial drift. Finery slag contained nearly as much iron as some ores. Little of it remains at forge sites today because the forge operators resmelted it in bloomery hearths or added it to the burden of the region's blast furnaces. Operators of rolling and slitting mills collected the scraps of metal from their machinery and furnaces for reworking at forges. Some of these practices lapsed in the prosperous 1880s, but when the profits from the blast furnaces began to decline, the New England Slag Company erected a plant to mine and crush accumulated slag into grit useful in roofing materials, road metal, and aggregate for concrete. In a terminal act of recycling, salvagers took the stone facings of abandoned blast furnaces to build bridge abutments and other structures.

The reliance of Salisbury ironmakers on oxidation and reduction reactions with carbon, silicon, and iron minimized their chemical impact on the environment. The principal solid waste, ironmaking slag, consisted of silicon, aluminum, and iron oxides. It was chemically inert once out of the forge hearth or furnace, and it released no toxic substances into the environment. Charcoal, unlike mineral coal, did not form clinkers or ash containing heavy metals.

Coaling wood released smoke bearing pyroligneous acid that could be a source of local air pollution. However, the dispersion of colliers' pits throughout the woodlands minimized the smoke nuisance, and so far no long-term environmental impact has been found. The distillation plant in East Canaan was a source of water pollution, but it operated for too short a time to have done long-term environmental damage. Thus, the reliance of the Salisbury ironmakers on renewable energy sources in place of the mineral coal that ironworks elsewhere increasingly used kept the district largely free of chemical degradation of its environment.

Comparisons with other ironmaking regions that enjoyed similar natural resources and markets could show us how Salisbury preferences rather than geographic factors particular to the region determined its industrial ecology. Immediately across the state line in New York, the Livingstons had furnaces and forges that used the same ore as those in Salisbury. The Livingstons ran them as part of a manor system that extracted inadequately rewarded labor from tenants. This master and servant relationship discouraged the cooperation among artisans and managers needed to make high-quality metal, and the Livingston forges never competed with those in Salisbury for the gun-iron and other high-end markets. Eventually, after the Livingstons abandoned their ironworks, new owners incorporated some of the New York blast furnaces into their Salisbury operations.

Ironmakers in northern New Jersey and in the Adirondack region of New York, active participants in the forge branch of the industry during the early Republic, had abundant water and forest resources, as well as magnetite ore that lent itself to bloomery smelting. Here, as in Salisbury, the charcoal-fired forges and furnaces left little long-term environmental impact. However, ironmakers in New Jersey and the Adirondacks took the opportunity offered them by cheap transportation on the Morris, Delaware and Hudson, and New York State Canals to go into large-scale mining of ore for export and pig-iron smelting with mineral-coal fuel. Abandoned mines, blast furnace sites, and, particularly in the Adirondacks, great piles of mine wastes remain on the land from these ventures.

Ironmakers in western Maryland, as at Antietam forge and furnace, and in the Shenandoah valley of Virginia had comparable resources, and built furnaces and forges similar in size and technique, to those in Salisbury. The Antietam ironworks was just a few miles up the Potomac River from the national armory at Harpers Ferry. The Antietam artisans failed to match the quality of the Salisbury product and so never captured the lucrative armory purchases of gun iron from its Connecticut competitors. The social barrier between master and slave, as documented by Charles Dew for the Tredegar works and Buffalo Forge in Virginia, eliminated the cooperative effort and free exchange of ideas between artisan and manager necessary to attain production of metal of the highest quality at these Southern works.[3] The forges left little environmental impact, but Tredegar's heavy use of mineral coal is still evident at its Richmond site.

Eastern Pennsylvania had charcoal-fired furnaces, such as the now-preserved Hopewell and Cornwall works, that supplied neighboring finery forges, such as Valley Forge. Ironmasters here adopted the indirect process (furnace and finery) thirty years before anyone did in Salisbury. Nevertheless, the Pennsylvania forges never became significant competitors in the particular market dominated by the Salisbury producers, despite having access to comparable natural resources. Many of the Pennsylvania forges operated as iron plantations, small villages in which absentee owners living in Philadelphia or Pittsburgh controlled the housing, land, and stores used by workers. These owners left management in the hands of agents or, sometimes, a partner who nevertheless retained close family connections in the city. Most of the Pennsylvania forge owners were interested in quantity production, and they shifted their capital to coal-fired furnaces and puddling works during the 1840s.

The forge branch was a relatively small component of ironmaking in other regions of the United States where smelting with charcoal throve: Alabama, Missouri, northern Michigan, Kentucky, Tennessee, and Ohio. Ohio's Hanging Rock district, a major producer of charcoal iron in the middle and late nineteenth century, had a number of blast furnaces controlled by Welsh immigrants who were also farmers. All the Hanging Rock furnaces produced merchant pig iron and lacked forges to make bar iron. The Welsh ownership, a form of community capitalism that lasted only about a decade, was unlike the continuous, locally controlled ownership of the Salisbury ironworks.[4]

Iron ore gave people in the Salisbury region opportunities to be part of one of the young nation's most important industries. Within the constraints of their particular geographic setting and the absence of mineral fuel in Connecticut, they had wide latitude to direct their industry along distinctively different, alternative paths. They made five key choices that defined the region's industrial ecology.

- They bought out the interests of non-resident investors within the first few decades of their industry and thereafter kept full control of the mines, furnaces, and forges within the district.
- People throughout the community participated in the industry by working, sometimes interchangeably, as owners, managers, and artisans.
- As soon as the nation's new manufacturing industries created a market for high-value, high-quality metal, Salisbury ironmakers elected to concentrate their production for this specialized, low-volume market rather than compete with other regions of the United States with tonnage production of ordinary grades of metal.
- After some modest experimentation, ironmakers in Salisbury eschewed mineral-coal fuel. Instead, they managed the region's woodlands for continuous production of wood fuel and used falling water for their mechanical power.

- Confronted with customer dissatisfaction as, with expanded production, the limitations of artisanal skill led to inadequate control of the quality of the metal they made, Salisbury's ironmakers chose to abandon ironmaking in favor of alternative enterprises rather than adopt new technique or learn about developments in metallurgical science.

The history of Salisbury shows that while natural resources, topography, and transportation systems all had their influence on the course of industry and its environmental consequences, components of the region's particular culture played a vital role in its industrial ecology. Accumulation of wealth and power had a weaker hold on people here than it did elsewhere. Appreciation of the natural beauties of the region, dislike of city capitalists, the respect accorded both skilled artisans and the professions, and limited interest in technological advances defined the cultural component of Salisbury's industrial ecology. Combined with adherence to proven technique and concentration on high-value, low-volume products, Salisbury's particular culture helped the region avoid the heavy environmental and social costs that typically accompanied extractive industries elsewhere in the United States.

APPENDIX ONE

Ironworks Inventory

Charles Rufus Harte, H. C. Keith, and, recently, Edward Kirby located the sites of all the blast furnaces in the Salisbury district. They established the dates when they were built and abandoned, and in some cases when they were converted to hot blast.[1] These data are summarized in table A1.1, and the furnace locations shown in figure A1.1.

No bloomery or finery forge or puddling works in the Salisbury district has yet been excavated to professional standards by archaeologists. The documentary record of the bloomery forges is particularly weak but probably could be improved by further study of land records. Construction dates for many of the bloomery forges are known only to within a decade, and few dates of abandonment are known (table A1.2). The record for the finery forges (table A1.3) is significantly better because of their larger size and sales to the federal government. The use of the puddling process in Salisbury seems to have been completely forgotten until our research on the Ames ironworks.[2] There is no secondary literature on them, and exact dates over which they operated remain to be discovered (table A1.4).[3]

Table A1.1. Blast Furnaces in the Salisbury District

Location	Built	HB	Closed	Years operated
1 Lakeville	1762	—	1832	70
2 Mount Riga	1810	—	1856	46
3 Canfield	1812	—	1857	45
4 Macedonia	1823	ca. 1847	1860	37
5 Sharon Valley	1825	ca. 1863	1898	73
6 Kent	1825	1846	1892	67
7 Bulls Bridge	1826	1844	1865	39
8 Chapinsville	1826	<1856	1897	71
9 Lime Rock No. 1	1830	<1856	1857	32
10 West Cornwall	1832	—	1857	25
11 Forbes (Canaan 1)	1832	1856	1880	48
12 Cornwall Bridge	1833	<1866	1897	64
13 New Preston	1837	—	1856	19
14 Scoville	1844	1853	1857	13

(Continued)

Table A1.1. Blast Furnaces in the Salisbury District (*continued*)

Location	Built	HB	Closed	Years operated
15 Weed's	1845	—	1866	21
16 Beckley	1846	1846	1919	82
17 Joyceville	1847	—	1854	7
18 Buena Vista	1847	1847	1893	46
19 Lime Rock No. 2	1865	1865	1900	35
20 Canaan No. 3	1872	1872	1923	51
21 Canaan No. 4	1918	1918	1919	0

Notes: Furnace numbers correspond to those on the map (fig. A1.1). HB indicates year hot blast was added. Furnaces may have been in blast intermittently during some of the years operated. Based on data from published sources, manuscripts, and land records. I thank Edward Kirby for the data on the Weed and Joyceville furnaces.

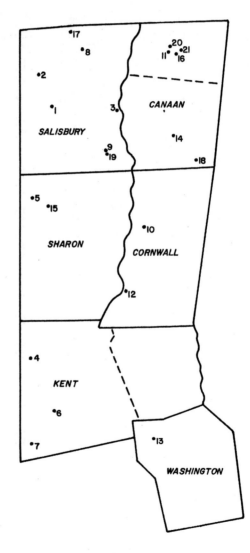

Fig. A1.1. *Map showing the locations of blast furnaces in the Salisbury district. Numbers correspond to those in table A1.1.*

Table A1.2. Bloomery Forges in the Salisbury District

Built	Closed	Location	Owners or Name
1733	1792	Litchfield	Kellogg; Rowe
1733	—	New Milford	Still River Forge
1734	1837	Lime Rock	Lamb; Holley & Coffing
1739	1850	Canaan	Seymour; Forbes; Forbes & Adam
1740	—	Sharon	Skinner
1743	—	Canaan	Hosford; Burrall; Hunt; Coe
1744	1860	East Kent	Barnum; Morgan
1745	1800	New Preston	Cogswell
1748	—	Chapinsville	Bacon & Parke; Camp; Chapin
1748	—	Lakeville	Williams, Stoddard & Spencer; Smith
1753	—	Milton	Welch
1760	—	Norfolk	Forbes; Day
1766	1826	Bulls Bridge	Bulls
1768	—	New Milford	Nicholson
1770	—	Macedonia	Wilson
1779	1863	Canaan	Hunt
1780	1803	Cornwall	Sedgwick
1781	—	Mount Riga	Woodin; Ball
1781	—	Milton	Simons
1783	1850	Woodville	Hitchcock & Pratt; Adam
1794	1850	Woodville	Guthrie; Adam
1796	—	Macedonia	Converse
1799	1850	Woodville	Peters; Adam
179x	—	South Kent	Carter; Fuller
179x	—	Washington	Bantam River Forge
179x	—	Canaan	Burtt
179x	—	Canaan	Huntington & Day
17xx	—	Canaan	Hanchett
179x	—	Torrington	Torrington Forge
179x	—	Goshen	Goshen Forge
180x	—	Norfolk	Phelps
180x	—	Canaan	Reed Brook Forge
180x	—	Cornwall	Potter Brook Forge
180x	—	Cornwall	Furnace Brook Forge
180x	—	East Kent	West Aspetuck Forge
180x	—	Kent Falls	Kent Falls Forge
185x	—	Sharon	Weed & Cray

Notes: Most dates are approximate. No closure dates have been found for most of these forges. Compiled from maps, manuscript and published sources, and field observations. This list is probably incomplete.

Table A1.3. Finery Forges in the Salisbury District

Built	Closed	Location	Owners or Name
1770	1810	Robertsville	Smith; Ogden; Burr; Beeman; Squire
1789	1803	Colebroook	Rockwell Upper and Lower
1795	1850	Winsted	Jenkins & Boyd
1803	1853	Winsted	Rockwell No. 1
1808	1845	Winsted	Boyd
1808	1845	Winsted	Rockwell No. 2
1810	1850	Mount Riga	Holley, Coffing & Pettee Upper Forge
1810	1850	Mount Riga	Holley, Coffing & Pettee Lower Forge
1811	1850	Winsted	Cook
1825	1858	Falls Village	Canfield & Sterling, Robbins
1828	1857	Lime Rock	Holley & Coffing; Canfield & Robbins
1832	1859	Salisbury	Salisbury Center Forge
183x	1863	Canaan	Hunt
1834	1850	Canaan	Scoville & Church

Notes: Most dates are approximate. Compiled from published and manuscript sources and from field observations.

Table A1.4. Puddling Works in the Salisbury District

Built	Closed	Location	Owners or Name
1832	1871	Amesville	Ames Iron Works
183x	—	Macedonia	Kent Puddling Works
1836	1850	Mount Riga	Salisbury Iron Company
184x	—	Milton	Welch
1847	1863	New Hartford	Camp & Mansfield
1847	—	West Norfolk	Lawrence & Day
1853	—	West Norfolk	B. N. Stephens
1861	1869	Huntsville	Hunt Canfield

Notes: This list is based on documentary and archaeological data that remain incomplete.

APPENDIX TWO

Ironmaking and Steelmaking Techniques

English and Continental artisans who came to British North America through colonial times applied the ironmaking techniques they had used at home to American resources. They made bar iron directly from ore in bloomery forges and smelted pig iron in blast furnaces. They could convert pig to bar iron with finery forges, and bar iron into blister steel with cementation furnaces.[1]

Bloomery

A bloom smelter reduced iron ore directly to solid wrought iron in a hearth with a fire burning charcoal fuel and blown with an air blast from bellows. He formed a bloom, a lump of solid, white-hot iron containing liquid slag, in the hearth. He then took the still-hot bloom to a water-powered helve hammer, where, with successively heavier blows, he forced as much of the entrained slag out of the metal as possible. A waterwheel to drive the bellows was a convenience but not a necessity. Because a bloomery could be built with modest capital and could be operated by one or two artisans, adventurers on the frontier usually began making iron with this technique.

We lack a documentary record of how eighteenth-century Salisbury artisans designed and operated their bloomery forges. Their technique was probably similar to that practiced at the Hotchkin-Snow forge in Madison, Connecticut, where the proprietors and their helpers made iron for the surrounding community from 1790 to 1819. Because this forge site remained undisturbed after smelting ended, it contains a physical record of the technique used. Hotchkin and Snow placed their bloomery hearth immediately behind the dam that formed the forge pond. They built their water-powered helve hammer adjacent to the hearth at the head of a tail race. The hearth, made of flat stones, had a working volume about twelve inches square in plan and about ten inches deep; it could have accommodated a bloom weighing up to about eighty pounds. It had a tap hole that the bloomer could use to draw off excess slag.[2]

Blast Furnace

A bloomery produced solid wrought iron—essentially pure iron with entrained slag particles—that a smith could forge into wrought products. To make hollowware and other cast products, iron had to be melted. A blast furnace made iron containing enough carbon and silicon so that it would melt at a relatively low temperature. Furnacemen could draw liquid metal from the furnace hearth in ladles to pour into molds to make pots and other cast products, or run it into long bars, called *pigs*, intended for subsequent remelting. Artisans fed ore, charcoal fuel, and limestone flux into the top of the furnace shaft. A pump called the *blast* or *blast engine* blew air into the bottom of the shaft. The countercurrent flow of air up and of ore, fuel, and flux down produced liquid alloy that collected in a pool covered with liquid slag in the crucible at the bottom of the furnace shaft. The seventeenth-century blast furnace design used in New England had a massive stone stack with two or more arches at the bottom to allow access to the hearth. Since the furnace ran continuously for weeks or months, a water-powered blast engine was always used, and a reliable water supply was a prime factor in selecting a furnace location.

The earliest Connecticut blast furnace stack to survive is the one on Mount Riga (see fig. 4.2), built in 1810. In addition to the tapping arch, it has a single blowing arch in a stack of rubblestone twenty-four feet square at the base and twenty-four feet high.[3] Its size and design are comparable to those of the 1647 Saugus furnace.[4] The northwest's first blast furnace, at Lakeville, would have been similar to that on Mount Riga. Until at least the 1820s, Connecticut blast furnace builders followed the seventeenth-century furnace design.

Operation of a blast furnace taxed the financial resources and technological skills of its proprietors, who had to organize a team of artisans and gather adequate stocks of ore, fuel, and flux to keep the furnace running for weeks or months without interruption. In 1776, when casting cannon for the Continental army, the Lakeville furnace had a staff of fifty-nine, of whom thirty were colliers and ore diggers; the rest ran the furnace and cast the cannon. A mistake that led to the furnace freezing up or a breakout of the molten iron could well bankrupt the furnace owners.

Finery Forge

Since a blast furnace ran continuously, it could make more metal than its proprietors could use for castings at any given time. Furnacemen cast the rest into pigs, which might be remelted in a cupola or air furnace at a foundry for pouring castings or be sold to a finery forge for conversion to wrought iron. In the northwest, most of the blast furnace output went to finery forges until the mid-nineteenth century.

A finer melted iron pigs in a charcoal fire with a strong air blast that burned the carbon and silicon out of the metal. He made a mass of solid iron and liquid slag similar to a bloom, and often called a *loup*. The finer hammered it, much as in a bloomery, to force out as much slag as possible and convert the loup into bar iron.

The proprietors of the Saugus ironworks introduced fining technique to America in 1647.[5] Salisbury artisans probably used the same technique when they began fining. Fining required particular skills, and forge proprietors frequently complained of the difficulty of recruiting competent finers.

It was cheaper to haul pig iron to a forge site than the amount of ore needed to make an equivalent amount of bar iron by bloom smelting. Forge operators would choose fining over bloom smelting, even though it required more fuel, when the distance to the nearest iron mine was great.[6] Additionally, skilled finers could make wrought iron more nearly free of slag inclusions and phosphorus than bloomer smelters could. Ironmasters in the northwest gradually shifted from the direct ironmaking process (bloom smelting) to the indirect process (blast furnace smelting and fining) as they concentrated on making the highest grade, most expensive bar iron.

Steel

Bar iron could be converted to steel by holding it at a high temperature in contact with carbon. Artisans packed iron bars in a chest with charcoal dust, then sealed the chest and heated it red-hot for about a week in a cementation furnace. The homogeneity of the product, known as *blister steel*, could be improved by breaking the bars into short lengths, stacking them, and welding them together under a hammer, making a product known as *shear steel*. If the blister steel were melted in crucibles and cast into ingots, it could be made free of included slag and more homogeneous than shear steel. No one in the United States mastered the technique of making *crucible steel* or *cast steel*, until the middle of the nineteenth century.

Fuel

Wood converted to charcoal served as the principal fuel for ironmaking in the northwest. After woodmen chopped logs into lengths of about four feet, colliers, usually working as independent contractors, undertook the coaling of the wood. A collier and his helper stacked the wood into a closely packed, circular pile, called a *pit*, that they covered with sod and charcoal dust. They set the pit alight and closely controlled the admission of air so that the fire would drive off the moisture and volatile constituents of the wood, leaving nearly pure, porous carbon behind. By the mid–nineteenth century, advanced ironworks began using kilns instead of pits; in the early twentieth century they used distillation retorts that recovered the products of combustion.

Because fuel was a major component of the cost of running ironmaking furnaces, efficient fuel use is an indication of the technological sophistication of the ironmakers. It is measured by the fuel rate, the ratio of the weight of charcoal used to the weight of iron made. Best-practice fuel ratios could be as low as 1.3 for a bloomery, 0.7 for a finery, and 0.6 for charcoal-fired blast furnaces.[7]

Wastes

Slag was the principal waste product of ironmaking. Because bloomery and finery slag contained enough iron to be worth resmelting, ironmasters in the northwest often recycled it. Hence, the quantity of slag remaining at forge sites is usually small.

Blast furnace slag contained little iron other than small particles of entrained metal. In the early years of the Salisbury district, finery forge operators near blast furnaces sometimes broke up the furnace slag in stamp mills to recover the metal shot it contained. Otherwise, furnace proprietors simply dumped their slag on the nearest land they did not want for other purposes.

Units and Conversion Factors

In the early Republic, account books were kept in both dollars and sterling. When conversions were made, they were at the rate of £1 equals $3.33.

Salisbury ironworkers usually recorded quantities of iron and ore in units of tons, hundredweight, quarters, and pounds. In the most commonly used system, a ton contained 20 hundredweight; each hundredweight contained four quarters; and each quarter, 28 pounds. However, some transactions used short tons (2,000 pounds instead of 2,240 pounds), pig tons (a ton plus a quarter to allow for sand adhering to the pigs), or bloom tons (twenty-two hundredweights). Documents often fail to specify the unit used.

In Salisbury, quantities of charcoal were usually given in bushels, or loads of a hundred bushels. All the calculations of fuel use and wood consumption in this book are based on 16 pounds as the average weight of a bushel of charcoal, a yield on coaling of thirty-three bushels from a cord of wood, and a forest yield of twenty cords per acre.[1]

Notes

PREFACE

1. Detailed explanations of ironmaking processes are found in Robert B. Gordon, *American Iron, 1607–1900*, Baltimore, Johns Hopkins University Press, 1996.

2. Robert B. Gordon and Michael S. Raber, *Industrial Heritage in Northwestern Connecticut, a Guide to History and Archaeology*, New Haven: Connecticut Academy of Arts and Sciences, 2000.

1. INDUSTRIAL ECOLOGY IN HISTORICAL PERSPECTIVE

1. A colloquium organized by K. N. Patel and held at the National Academy of Sciences in May 1991 helped participants define industrial ecology. The papers presented were published in the *Proceedings of the National Academy of Sciences*, vol. 89 (1992). For an introduction see T. E. Graedel and B. R. Allenby, *Industrial Ecology*, Englewood Cliffs, N.J.: Prentice Hall, 1995. Current research in the field appears in the journal *Industrial Ecology*, launched in 1997.

2. B. J. Skinner, "A Second Iron Age Ahead," *American Scientist* 64 (1976): 258–269; Skinner, "Earth Resources," *Proceedings of the National Academy of Sciences* 76 (1979): 4212–4217.

3. The district included the towns of Canaan, Salisbury, Sharon, Cornwall, Kent, and Roxbury (which had blast furnaces) and Colebrook, Norfolk, Winchester, Litchfield, Warren, Washington, and New Milford (which had forges and rolling mills). Later, district ironmakers also owned furnaces and mines in the Berkshire hills of Massachusetts as far north as Adams, and in immediately adjacent New York counties. At times some of them imported fuel from Vermont. "Salisbury" is used in this book to designate the entire district, unless the town itself is specified. Descriptions of the district towns can be found in John W. Barber, *Connecticut Historical Collections*, New Haven: Durrie, Pack and Barber, 1838. Sites of ironworks are shown by Moses Warren and William Gillet, *Connecticut from Actual Survey Made 1811*, Hartford, 1812; William Blodget, *A New and Correct Map of Connecticut*, Middletown, 1792; *Clark's Map of Litchfield County*, surveyed by G. M. Hopkins Jr., Philadelphia: Richard Clark, 1859; *Geographical Location of Historical Landmarks in the Town of Salisbury, Connecticut*, Salisbury: Salisbury Association, 1949; *Topographical and Historical Map of the Town and Villages of Salisbury*, topography by I. W. and E. F. San-

ford, historical sketches by Malcolm D. Rudd and I. W. Sanford, New York: Sanford, 1899.

4. Robert B. Gordon and Patrick M. Malone, *The Texture of Industry*, New York: Oxford University Press, 1994, chaps. 5 and 7.

2. RESOURCES DISCOVERED

1. James A. Mulholland, *A History of Metals in Colonial America*, Tuscaloosa: University of Alabama Press, 1981, pp. 62–116.

2. Julia Pettee, *The Rev. Jonathan Lee and His Eighteenth Century Parish: The Early History of the Town of Salisbury, Connecticut*, Salisbury: Salisbury Association, 1957, pp. 20–21.

3. James Morris, *Statistical Account of Several Towns in the County of Litchfield*, pp. 85–124 of *A Statistical Account of the Towns and Parishes of Connecticut*, New Haven: Connecticut Academy of Arts and Sciences, 1811, p. 86.

4, Jackson T. Main, *Society and Economy in Colonial Connecticut*, Princeton, N.J.: Princeton University Press, 1985, pp. 115, 123; Bruce C. Daniels, *The Connecticut Town*, Middletown: Wesleyan University Press, 1979, p. 140.

5. Samuel Church, "Historical Address" (delivered at the Centennial Town Meeting of Salisbury, 20 October 1841), in *Historical Addresses*, Pittsfield, 1876, p. 38; Pettee, p. 58.

6. Pettee, p. 61.

7. Historians have yet to identify the initial proprietors of the Kent ore bed.

8. Charles S. Grant, *Democracy in the Connecticut Frontier Town of Kent*, New York: Columbia University Press, 1961, pp. 10, 16–22; Douglas R. McManis, *Colonial New England: A Historical Geography*, New York: Oxford University Press, 1975, p. 65.

9. Church, pp. 38, 10, 11; Pettee, pp. 45–47, 58.

10. William L. Warren, "The Litchfield Iron Works in Bantam, Connecticut," *Tools and Technology*, July 1992, pp. 1–27.

11. Richard S. Allen, "Connecticut Iron and Steel from Black Sea Sands," *IA, Journal of the Society for Industrial Archeology* 18 (1992): 129–132.

12. David W. Crossman, "A New Year's Discourse Delivered at Salisbury . . . January 2nd 1803: Containing the Ancient History of the Town," Hartford: Hudson and Goodwin, 1803, p. 11; Morris, p. 86; Lewis M. Norton, "Goshen," and Barzillai Slosson, "Kent," both in *Connecticut Towns, 1800–1832: Voices of the New Republic*, vol. 1, *What They Said*, ed. Howard Lamar and Carolyn Cooper, New Haven: Connecticut Academy of Arts and Sciences, 2000.

13. Crossman, p. 11.

14. D. H. Newland, "Mineralogy and Origin of the Taconic Limonites," *Economic Geology* 31 (1936): 133–155; E. Zen and J. H. Hartschorn, "Geologic Map of the Bashbish Quadrangle," Map GQ-507, Washington, D.C.: U.S. Geological Survey, 1966; R. M. Gates, *The Bedrock Geology of the Sharon Quadrangle*, State Geological and Natural History Survey of Connecticut, Quadrangle Report No. 38, 1979. Geologists were uncertain about the formation of the Salisbury ore through the nineteenth century. Charles U. Shepard (*A Report on the Geological Survey of Connecticut*, New Haven, 1837, p. 17) described the ores as fibrous brown hematite, compact hematite, and an ochery mixture of the two in beds with a stratification conformable with the mica-slate. J. D. Dana ("Berkshire Beds of Iron," *Berkshire Historical and Scientific Society Collection* 1

[1886]: 19–25) believed the ore was formed by weathering of calcium, magnesium, and iron carbonates, not the schist. William H. Hobbs ("The Iron Ores of the Salisbury District of Connecticut and Massachusetts," *Economic Geology* 2 [1907]: 153–181) believed that the ore was deposited in pockets formed by faults from solutions leached from the schist. His theory was repeated by J. F. Schairer. (Geologists today are unable to locate many of the faults Hobbs identified.) The ore at Kent is at the contact of Berkshire schist and Poughquag quartzite (J. F. Schairer, *The Minerals of Connecticut*, State Geological and Natural History Survey of Connecticut, Bulletin 51, 1931, p. 102).

15. Crossman, p. 10.

16. Isabel S. Mitchell, *Roads and Road-making in Colonial Connecticut*, New Haven: Yale University Press, 1933.

17. Percy W. Bidwell, "Rural Economy in New England at the Beginning of the Nineteenth Century," *Transactions of the Connecticut Academy of Arts and Sciences* 20 (1916): 241–399; see p. 313.

18. Philip E. Taylor, "The Turnpike Era in New England," Ph.D. diss., Yale University, 1934; Fredric J. Wood, *The Turnpikes of New England*, Boston: Jones, 1919; Robert B. Gordon, "Travel on Connecticut's Roads, Bridges, and Ferries, 1790–1830," in *Connecticut Towns, 1800–1832: Voices of the New Republic*, vol. 2, *What We Think*, ed. Howard Lamar and Carolyn Cooper, New Haven: Connecticut Academy of Arts and Sciences, 2000.

19. Quoted in Jarvis M. Morse, *A Neglected Period in Connecticut's History, 1818–1850*, New Haven: Yale University Press, 1933; reprint, New York: Octagon Books, 1978, p. 257.

20. Church, p. 15.

3. INDEPENDENT ARTISANS

1. William L. Warren, "The Litchfield Iron Works in Bantam, Connecticut," *Tools and Technology*, July 1992, pp 1–27.

2. *Salisbury Town Meeting Minutes*, ed. Geoffrey Rosanno, Salisbury: Salisbury Association, 1988, p. 147.

3. Samuel Church, "Historical Address" (delivered at the Centennial Town Meeting of Salisbury, 20 October 1841), in *Historical Addresses*, Pittsfield, 1876, p. 36; Julia Pettee, *The Rev. Jonathan Lee and His Eighteenth Century Parish: The Early History of the Town of Salisbury, Connecticut*, Salisbury: Salisbury Association, 1957, pp. 58, 60.

4. Church, p. 37; Pettee, p. 61.

5. Charles S. Grant, *Democracy in the Connecticut Frontier Town of Kent*, New York: Columbia University Press, 1961, p. 41.

6. Manuscript notes in Falls Village Historical Society include a copy of the Canaan Land Records of 26 August 1800: William and Elizabeth Burrall of Canaan conveyed half the forge (with its coal house, tools, and utensils, and the right to raise ore in Salisbury Ore Hill) and one-quarter of the gristmill standing below the Great Falls to their son William M. Burrall.

7. Emily M. Hopson, *Kent Tales: Iron Fever*, Kent, Conn.: Kent Historical Society, 1990, p. 2.

8. Ibid., pp. 3–8.

9. Kenneth Howell and Einar W. Carlson, *Empire over the Dam*, Chester, Conn.: Pequot Press, 1974, pp. 131–132.

10. Church, p. 37.

11. Subsequent proprietors included Caleb Smith, John Dean, Gideon Skinner, Joseph Jones, Eliphalet Owen, John Cobb, and Leonard Owen (ibid.).

12. Ibid.; Pettee, pp. 114, 134, 169.

13. Samuel Orcutt, *History of the Towns of New Milford and Bridgewater*, Hartford: Case, Lockwood and Brainard, 1882, p. 364; Warren, p. 18.

14. Kenneth Howell and Einar W. Carlson, *Men of Iron: Forbes and Adam*, Lakeville, Conn.: Pocketknife Press, 1980, pp. 25–31.

15. Ibid., p. 69.

16. Emily M. Hopson, *Kent Tales: Rufus Fuller and the South Kent Ore Bed*, Kent, Conn.: Kent Historical Society, 1988, pp. 4–5; Howell and Carlson, *Men of Iron*, p. 75; Howell and Carlson, *Empire over the Dam*, p. 109.

17. Hopson, *Kent Tales: Rufus Fuller*, p. 40.

18. On these trading networks, see, for example, Christopher Clark, *The Roots of Rural Capitalism*, Ithaca, N.Y.: Cornell University Press, 1990, pp. 33–37, 65–67.

19. "Simeon Palmer's Trading," ms. at the Litchfield Historical Society.

20. Robert B. Gordon, "Material Evidence of Ironmaking Technique," *Journal of the Society for Industrial Archeology* 21 (1995): 69–80; Gordon, "Process Deduced from Ironmaking Wastes and Artifacts," *Journal of Archaeological Science* 24 (1997): 9–18.

21. Gordon, "Material Evidence."

4. MERCHANT CAPITALISTS

1. This was the cost of the first blast furnace built in Salisbury. For the bloomery estimate, see J. B. Bailey to George Throop, 26 August 1848, Bailey papers, Clinton County (New York) Historical Museum.

2. William F. Morrill, "The Spur at the Head of the Manor, Salisbury, Connecticut, and the Livingstons," in *The Livingston Legacy*, ed. Richard T. Wiles, Annandale, N.Y.: Bard College, 1987, pp. 38–66.

3. I thank Ronald Jones, a descendant of the Hazeltines, for the information about John and Paul.

4. Contract in the possession of Dr. William Adam.

5. A drawing of the furnace prepared in 1878 from unknown sources shows a furnace twenty-eight feet high with a nine-foot greatest inside diameter and four arches. It had a capacity of 2.5 tons per day and operated at a fuel ratio of 1.8. The furnace builders used "firestone" from Haverstraw on the Hudson for the crucible, "slate" for the interior of the shaft, yellow sand for the lining between the inwall and the outer structure, and limestone blocks from Lakeville quarries for the outer walls. (Anon., "Salisbury Iron, Its Composition, Qualities, and Uses," reprinted from the *New York Railroad Gazette*, Salisbury: Barnum, Richardson Co., 1878.)

6. Furnace staff based on data in T. Pownall, *A Topographical Description of the Dominion of the United States of America*, Pittsburgh: University of Pittsburgh Press, 1949, p. 67.

7. Samuel Church, "Historical Address" (delivered at the Centennial Town Meeting of Salisbury, 20 October 1841), in *Historical Addresses*, Pittsfield, 1876, p. 37; Kenneth Howell and Einar W. Carlson, *Men of Iron: Forbes and Adam*, Lakeville, Conn.: Pocketknife Press, 1980, pp. 37–38; Julia Pettee, *The Rev. Jonathan Lee and His Eighteenth Century Parish: The Early History of the Town of*

Salisbury, Connecticut, Salisbury: Salisbury Association, 1957, pp. 114, 127, 134, 169. In 1776 a staff of fifty-nine, of whom thirty were colliers and ore diggers, ran the furnace and cast cannon.

8. Paul Hazeltine left Salisbury to join his father in Townshend, Vermont, and Allen left for Northampton, Massachusetts, where two years later the citizens, dismayed by his profane ways, asked him to leave (Pettee, p. 131). Since Hazeltine sold his half interest for £1,530, it appears the partners made an investment of £2,620 in addition to their initial purchase to build the blast furnace plant.

9. Pettee, pp. 127, 132, 136–138, 169. Smith also acquired the shares held by Elisha Forbes's estate in 1769. Documents describing the Caldwells' partnership agreement with Smith, the accumulated debts, and the actions of the General Assembly are found in the Richard Smith Business Papers at the Baker Library, Harvard University Business School.

10. Lane to Smith, 7 October 1773, Smith Business Papers, Folder 17.

11. Irving E. Manchester, *The History of Colebrook*, Colebrook, Conn.: Sesqui-centennial Committee, 1935, pp. 36–44; Pettee, pp. 154, 170; Anon., *History of Litchfield County, Connecticut*, Philadelphia: Lewis, 1881, p. 277.

12. Smith paid Stephen Goller of New Hartford (then the closest village to Robertsville) in 1768 for "mending the dam at the forge" (Smith Business Papers, Folder 9). This may have been a bloomery forge that Smith converted to a finery once he had a supply of pig iron available.

13. Manchester, pp. 37–38. Smith's copy of the agreement reads, "Voted that Richard Smith of Boston shall have the Liberty to Draw off 1 1/2 foot of the Long Pond on Winchester for his Iron works (agreeing with the owners of the Land he shall cross) During the pleasure of the Proprietors" (Smith Business Papers, Folder 27).

14. Theodore Kury, paper presented at the 1996 Ironmasters conference.

15. Harry Miller, "Potash from Woodashes: Frontier Technology in Canada and the United States," *Technology and Culture* 21 (1980): 187–208. (See section on casting pots in Salisbury.)

16. Smith to Tilley, 15 February 1775, Smith Business Papers, folder 20.

17. Some of these suppliers are named in Louis F. Middlebrook, *Connecticut's Cannon*, Salem, Mass.: Newcomb Gauss, 1935, p. 17.

18. Middlebrook, pp. 14, 16, 24, 41; Pettee, pp. 180–181.

19. Pettee, pp. 170–173, 184–186.

20. Whiting and Nelson used the 1776 technique: they ran iron tapped from the furnace into clay molds without cores placed in casting pits. They then bored out the solid castings. (Manuscript notes by Malcolm Rudd, Scoville Memorial Library, Salisbury; Holley & Coffing to William Bainbridge, 1 November 1813.)

21. Pettee, pp. 170–173, 184.

22. Two blotters (1971-99-24 and 1971-99-27) and a ledger (1917-99-5) are at the Litchfield Historical Society. There is no inscription on the blotters or the ledger to identify the organization they belonged to. However, the ledger contains a number of loose documents for transactions made by Coffing & Holley that match entries in the ledger and in the blotters. Additional details on furnace operation are from Benjamin Silliman, "Sketches of a Tour in the Counties of New Haven and Litchfield in Connecticut," *American Journal of Science* 2 (1820): 201–235.

23. List of shipments, Smith Business Papers, folder 24.

24. M. Rudd, manuscript notes, Scoville Memorial Library, Salisbury.

25. Flask castings sold for $4.00 per cwt; open castings for $3.50 per cwt. (Holley & Coffing blotter); J. M. Holley and J. C. Coffing contract 2 May 1815. I thank Federick Chesson for allowing me to see this document.

26. Holley & Coffing to W. Dunbar, 12 March 1810, letter no. 702, Salisbury Association collection.

27. M. Rudd manuscript notes, Scoville Memorial Library, Salisbury; Holley & Coffing to Commodore William Bainbridge, 1 November 1813. I thank Federick Chesson for allowing me to see this letter.

28. Felicia J. Deyrup, *Arms Makers of the Connecticut Valley*, Northampton, Mass.: Smith College, 1948, p. 73.

29. Church, p. 38; Malcolm D. Rudd, *Men of Worth of Salisbury Birth*, Salisbury: Salisbury Association, 1991, pp. 200–201.

30. Letter in Coffing & Holley account book, Litchfield Historical Society.

31. Church, p. 38; Kenneth Howell and Einar W. Carlson, *Men of Iron: Forbes and Adam*, Lakeville, Conn.: Pocketknife Press, 1980, p. 119.

32. Church, p. 38. No records of how Bradley financed the furnace have been found.

33. Silliman, p. 210

34. Charles U. Shepard, *A Report on the Geological Survey of Connecticut*, New Haven, 1837, pp. 19, 25.

35. Silliman, p. 212.

36. Most of these furnaces were later enlarged and fitted with hot blast. Only the Mount Riga furnace remains as an example of early Salisbury furnace construction practice. (Robert B. Gordon and Michael S. Raber, *Industrial Heritage in Northwestern Connecticut*, New Haven: Connecticut Academy of Arts and Sciences, 2000.

5. ARTISAN-ENTREPRENEURS

1. Kenneth Howell and Einar W. Carlson, *Men of Iron: Forbes and Adam*, Lakeville, Conn.: Pocketknife Press, 1980, p. 36.

2. Elisha Forbes had been killed by a horse in 1765, leaving his brother to carry on the business.

3. Howell and Carlson, *Men of Iron*, pp. 43–45.

4. Theodore Kury, paper presented at the 1996 Ironmasters conference.

5. E. Wadsworth & Co. account book, slitting mill entries for 1798, Litchfield Historical Society ms. no. 1972-11-2B.

6. Howell and Carlson, *Men of Iron*, p. 70; William L. Warren, "The Litchfield Iron Works in Bantam, Connecticut," *Tools and Technology*, July 1992, pp. 1–27, see pp. 9, 18.

7. Bakewell to Adam, 24 November 1807, letter in possession of Dr. William Adam.

8. Howell and Carlson, *Men of Iron*, p. 78.

9. Forbes & Adam account book marked "Grist Mill No. 2" on cover, at the Falls Village—Canaan Historical Society.

10. Leman Church to Samuel F. Adam, 17 November 1832, letter in possession of Dr. W. S. Adam.

11. Connecticut Historical Society ms. no. 74341.

12. Ms. notes in Falls Village—Canaan Historical Society; Benjamin Silli-

man, "Sketches of a Tour in the Counties of New Haven and Litchfield in Connecticut," *American Journal of Science* 2 (1820): 201–235.

13. Richard J. Purcell, *Connecticut in Transition*, 1918; republished, Middletown: Wesleyan University Press, 1963, pp. 65–66, 74, 78–84.

14. Whitney visited the East Canaan forge in the spring of 1799. In a letter of 12 March 1799 he also asked the partners if they could recommend two nailers whom he could hire to forge some of the lighter parts of muskets. (Whitney to Forbes & Adam, letter in possession of Dr. W. Adam.)

15. Silliman's plates were to be 13 x 10 x 1/2 inch, made of either bloomed or refined iron. (Howell and Carlson, *Men of Iron*, p. 92.) J. & J. Townsend, Columbian Foundry, to Holley & Coffing, 26 February 1818; Roswell Lee to Holley & Coffing, 20 August 1818 and 30 December 1818. I thank Frederick Chesson for allowing me to see these letters.

16. Robert B. Gordon, "Choice of Method for Making Wrought Iron in the Salisbury District of Connecticut," *Journal of the Historical Metallurgy Society* 31, no. 1 (1997): 25–31.

17. Robert B. Gordon and Geoffrey Tweedale, "Pioneering in Steelmaking at the Collins Axe Factory," *Journal of the Historical Metallurgy Society* 24 (1990): 1–11.

18. Francis W. Rockwell, *The Rockwell Family* (privately published), Pittsfield, Mass., 1924, p. 68.

19. Daniel P. Tyler, *Statistics of the Condition and Products of Certain Branches of Industry in Connecticut*, Hartford: Boswell, 1846.

20. Anon., *History of Litchfield County, Connecticut*, Philadelphia: Lewis, 1881, p. 201.

21. Irving E. Manchester, *The History of Colebrook*, Colebrook, Conn.: Sesqui-centennial Committee, 1935. A similar case in New Milford led to a trial in Litchfield County Court, with the dam breakers being found guilty of trespass. (David Daggett, "A Brief Account of a Trial at Law . . . ," *Memoirs of the Connecticut Academy of Arts and Sciences* 1, no. 1 (1810): 131–134.)

22. Manchester, pp. 54–56; Anon., *History of Litchfield County, Connecticut*, p. 201; John Boyd, *Annals and Records of Winchester, Conn.*, Hartford: Case, Lockwood and Brainard, 1873; Lee to Rockwell, 18 July 1816, Springfield Armory records, Springfield, Massachusetts.

23. Manuscript notes by Malcolm Rudd, Scoville Memorial Library, Salisbury.

24. Felicia J. Deyrup, *Arms Makers of the Connecticut Valley*, Northampton, Mass.: Smith College, 1948, p. 74.

25. Howell and Carlson, *Men of Iron*, p. 79.

26. A. H. Holley Diary, ms. held by the Salisbury Association.

27. J. M. Holley to M. A. Holley, 1 and 11 January 1829, letters 716 and 717 held by the Salisbury Association.

28. Peter B. Porter to John Milton Holley, 8 August 1800. I thank Frederick Chesson for allowing me to see this letter.

29. "Furnace Notes," a ms. Found in the minute book of the Bachelor's Club by M. D. Rudd and now in the possession of Frederick Chesson.

30. On Hall and Moor see M. R. Smith, *Harpers Ferry Armory and the New Technology*, Ithaca, N.Y.: Cornell University Press, 1977, pp. 184–218, 278.

31. J. M. Holley to M. A. Holley, 22 September 1831, letter No. 771, Salisbury Association collection.

32. A. H. Holley Diary, Salisbury Association collection.

33. "Salisbury Temperance Society," Box 2, Holley papers, Connecticut Historical Society.

34. J. M. Holley to Lewis Cass, 2 January 1832; J. M. Holley to Myron Holley, 28 April 1834; Gibson Grayson & Co. to J. M. Holley, 17 August 1833. I thank Frederick Chesson for allowing me to see these documents.

35. This and the following paragraphs are based on a sample of 150 letters in the Landon, Moore & Co. file of the Barnum-Richardson collection, Series V, Boxes 3–22, Archives and Special Collections, University of Connecticut at Storrs.

36. Lee to Holley & Coffing, 7 and 23 February 1831, Springfield Armory records.

37. Samuel Colt's diary, Colt papers, Connecticut Historical Society.

38. Order from Harpers Ferry Armory to Landon, Moore & Co., 21 December 1837, Barnum-Richardson papers.

39. Letters dated 14 June, 27 June, 11 July, and 15 July, Barnum-Richardson papers.

40. Letters received 22 May 1837, 28 February 1838, 28 February 1840, Landon, Moore & Co., Barnum-Richardson collection.

41. Ames to Landon, Moore & Co., 25 June 1844.

42. Canfield & Robbins to Ripley, 28 October 1845, Springfield Armory papers, Housatonic Railroad bills to Canfield & Robbins for 1844–1845 and 1849. I thank Frederick Chesson for allowing me to see these documents.

43. Smith to Moore, 23 January 1853; Harpers Ferry to Moore, 25 April 1850, 1 February 1853; Peck & Smith to Moore, 28 April 1852; Springfield Armory to Moore, 17 April 1852; Wilcox to Moore, 1 February 1853, Barnum-Richardson collection.

44. J. S. Whitney to Craig, 15 March 1855, Springfield Armory collection; Deyrup, pp. 136–138.

45. J. S. Whitney to S. B. Moore & Co., 12 December 1854, 26 December 1855, Barnum-Richardson collection.

46. Robert B. Gordon, *American Iron, 1607–1900*, Baltimore: Johns Hopkins University Press, 1996, pp. 129–133.

47. Robert B. Gordon, "Strength and Structure of Wrought Iron," *Archeomaterials* 2 (1988): 109–137.

48. Robert B. Gordon, "Materials for Manufacturing: The Response of the Connecticut Iron Industry to Limited Resources and Technological Change," *Technology and Culture* 24 (1983): 602–634; Gordon, "English Iron for American Arms: Laboratory Evidence on the Source of Iron Used at the Springfield Armory in 1860," *Journal of the Historical Metallurgy Society* 17 (1983): 91–98.

6. ENVIRONMENT, TECHNOLOGY, AND COMMUNITY IN SALISBURY

1. Anon., "Salisbury Iron," pamphlet published by Barnum & Richardson, 1878. Calculations of wood consumption here are based on yields of 20 cords of wood per acre and 33 bushels of charcoal per cord of wood (Herbert H. Winer, "History of the Great Mountain Forest, Litchfield County, Connecticut," Ph.D. dissertation, Yale University, 1955). These are average figures for the region, and varied for different areas of forest and coaling techniques.

2. John C. Smith, "Sharon," in *Connecticut Towns, 1800–1832: Voices of the New*

Republic, vol. 1, *What They Said*, ed. Howard Lamar and Carolyn Cooper, New Haven: Connecticut Academy of Arts and Sciences, 2000.

3. Contract dated 18 November 1796 in the possession of Dr. W. Adam.

4. These calculations assume that the blast furnace fuel rate was 1.8 (as it had been in 1779) through 1812 and 1.5 thereafter (by 1841 it was 1.4 [Samuel Church, "Historical Address" in *Historical Addresses*, Pittsfield, Mass., 1876]); that the finery-forge fuel rate was 2.7 (the rate known to have been achieved by 1845), and that all the bar iron was made by fining pig. Data for 1829 are from a census form filled out by J. M. Holley in 1832 (Scoville Memorial Library, Salisbury).

5. Malcolm D. Rudd, *Men of Worth of Salisbury Birth*, Salisbury, Conn.: Salisbury Association, 1991, p. 203.

6. Carolyn Merchant, *Ecological Revolutions, Nature, Gender, and Science in New England*, Chapel Hill: University of North Carolina Press, 1989, p. 157.

7. Robert B. Gordon, "Hydrological Science and the Development of Water Power for Manufacturing," *Technology and Culture* 26 (1985): 204–235.

8. Robert M. Thorson et al., "Colonial Impacts to Wetlands in Lebanon, Connecticut," ch. 3 in *Reviews in Engineering Geology*, vol. 12, ed. C. W. Webb and M. E. Gowan, Boulder: Geological Society of America, 1998.

9. Barzillai Slosson, "Kent," in *Connecticut Towns, 1800–1832: Voices of the New Republic*, vol. 1, *What They Said*, ed. Howard Lamar and Carolyn Cooper, New Haven: Connecticut Academy of Arts and Sciences, 2000.

10. Benjamin Silliman, "Sketches of a Tour in the Counties of New Haven and Litchfield in Connecticut," *American Journal of Science* 2 (1820): 201–235.

11. Charles U. Shepard, *A Report on the Geological Survey of Connecticut*, New Haven, 1837, p. 25.

12. "Forbes & Adam Washington Mill Ledger," account book at the Falls Village–Canaan Historical Society.

13. Silliman, p. 212.

14. Emily M. Hopson, *Kent Tales: Iron Fever*, Kent, Conn.: Kent Historical Society, 1990, p. 22.

15. Letter from Joseph Pettee, 31 October 1834, quoted by Julia Pettee, ms. Scoville Memorial Library.

16. C. and E. Hunt, and J. Boyle, to Landon, Moore & Co., 2 and 10 February 1841, Barnum-Richardson collection, University of Connecticut, Storrs.

17. Anon, *History of Litchfield County, Connecticut*, Philadelphia: Lewis, 1881, pp. 200–204.

18. Most Salisbury towns had textile mills, but they formed only a component of a diverse economy. Consequently, Salisbury did not suffer the conflict associated with rural capitalism in the mill towns of Massachusetts and Rhode Island. (Jonathan Prude, "Town-Factory Conflicts in Antebellum Rural Massachusetts," in *The Countryside in the Age of Capitalist Transformations*, ed. S. Hahn and J. Prude, Chapel Hill: University of North Carolina Press, 1985.)

19. Judith McGaw, *Most Wonderful Machine: Mechanization and Social Change in Berkshire Paper Making, 1801–1885*, Princeton, N.J.: Princeton University Press, 1987.

20. William Belcher to Oliver Ames, 22 March 1844, Arnold B. Tofias Industrial Archives, Stonehill College.

21. Holley & Coffing to Lee, 20 December 1820, Springfield Armory papers.

22. "Forbes & Adam's Book," also marked "Grist Mill No. 2," Falls Village-Canaan Historical Society; see p. 197.

23. Colliers individually negotiated contracts with the Cornwall Bridge Iron Company until the furnace closed in 1897. Examples of these contracts are found in Connecticut Historical Society ms. no. 74153.

24. Ms. History of Colebrook, Rockwell papers, box I, Connecticut Historical Society.

25. Gregory Galer, Robert Gordon, and Frances Kemmish, *Connecticut's Ames Iron Works*, New Haven: Connecticut Academy of Arts and Sciences, 1998.

26. *Litchfield Enquirer,* 20 June 1872.

27. *Litchfield Enquirer,* 8 July 1869.

28. *Litchfield Enquirer,* 2 December 1869.

29. Kenneth Howell and Einar W. Carlson, *Men of Iron: Forbes and Adam,* Lakeville, Conn.: Pocketknife Press, 1980, p. 70. Palmer's work is recorded in the E. Wadsworth & Co. account book at the Litchfield Historical Society.

30. See "Notes of things to be done when travelling," in the Canaan folder, Forbes and Adam papers, Rockwell papers, ms. no. 69460, Connecticut Historical Society.

31. M. D. Rudd ms. notes at the Scoville Library, Salisbury. On 1 September 1832 John M. Holley leased the Lime Rock works to his sons Alexander H. Holley, Francis N. Holley, and George W. Holley, along with John W. Calkins, who did business as Holley & Co. John M. Holley's widow, Mary Ann, sold the works to Canfield & Robbins on 20 February 1837. I thank Frederick Chesson for allowing me to see these contracts.

32. Samuel Forbes's blotter, 1827–1828, at the Falls Village–Canaan Historical Society.

33. Timothy Dwight, *Travels in New England and New York,* 4 vols., New Haven: T. Dwight, 1821, vol. 3, p. 261.

34. Forbes & Adam Washington Mills ledger, p. 73, at the Falls Village–Canaan Historical Society; ledger, "Ore Sent from Davis Mine to Adam & Beckley," Connecticut Historical Society ms. no. 74341.

35. Rudd, "Men," p. 170.

36. I thank Edward Kirby for the information about Weed.

37. Rudd, *Men of Worth,* p. 301; H. P. Harris, "Charcoal Manufacture in the Salisbury Iron Region," *Journal of the United States Association of Charcoal Ironworkers* 6 (1885): 49–53.

38. The marriages were Harriet to W. P. Burrall, Sally to S. S. Robbins, and Mary Ann to Moses Lyman (Rudd, "Men," p. 209). (Two other daughters died at a young age.)

39. A. H. Holley's autobiographical notes, Connecticut Historical Society.

40. Rudd, "Men," p. 5.

41. A. H. Holley's diary, Salisbury Association collection.

42. Samuel Church, "Historical Address" (delivered at the Centennial Town Meeting of Salisbury, 20 October 1841), in *Historical Addresses,* Pittsfield, 1876, p. 27. Bingham, born in Salisbury in 1757, went to an academy in Mansfield, Connecticut, completed his M.A. at Dartmouth in 1782, and became a Boston educator (Rudd, "Men"). E. S. Bartlett, "Salisbury," (*Connecticut Quarterly* 4 [1898]): 345–371, see p. 365.

43. Peter B. Porter to John M. Holley, 8 August 1800; Howell and Carlson, "Men of Iron" p. 51.

44. Rudd, "Men," p. 81.

45. Bruce C. Daniels, *The Connecticut Town*, Middletown: Wesleyan University Press, 1979, pp. 108–110; Julia Pettee, *The Rev. Jonathan Lee and His Eighteenth Century Parish: The Early History of the Town of Salisbury, Connecticut*, Salisbury: Salisbury Association, 1957, p. 116. Next to the appointment of town officials, the topic most frequently discussed at the Salisbury town meeting in colonial times was schools. (*Salisbury Town Meeting Minutes, 1741–1784*, ed. Geoffrey Rossano, Salisbury: Salisbury Association, 1988.)

46. Anon., *History of Litchfield County, Connecticut*, p. 550; Rudd, "Men," p. 209; Probate inventory, Salisbury town records.

47. Rudd, "Men," pp. 111, 164.

48. Jeanne McHugh, *Alexander Holley and the Makers of Steel*, Baltimore: Johns Hopkins University Press, 1980, pp. 12–19.

49. *Litchfield Enquirer*, 14 November 1867.

50. E. A. Ekern, "The Falls Village Hydroelectric Development of the Connecticut Power Co.," *Connecticut Society of Civil Engineers Annual Report* (1913): 110–130; Winton B. Rogers, "Short History of the Water Power Development at Falls Village," *The Lure of the Litchfield Hills* 8 (1951): 301, 315, 332. Because few primary documents from the Water Power Company have been found, its affairs remain somewhat mysterious. For a description of the substantial remains at Falls Village, see Robert B. Gordon and Michael S. Raber, *Industrial Heritage in Northwestern Connecticut*, New Haven: Connecticut Academy of Arts and Sciences, 2000.

51. Anon., *History of Litchfield County, Connecticut*, p. 554.

52. "Amount of Dividends Made on Ore Sold," ms. Coffing Estate Box, Connecticut Historical Society.

7. THE CHALLENGE OF NEW MARKETS AND TECHNIQUES

1. Julia Pettee, *The Rev. Jonathan Lee and His Eighteenth Century Parish: The Early History of the Town of Salisbury, Connecticut*, Salisbury: Salisbury Association, 1957, p. 154.

2. Samuel Orcutt, *History of the Towns of New Milford and Bridgewater*, Hartford: Case, Lockwood and Brainard, 1882, pp. 455–457; Charles R. Harte, "Connecticut's Canals," *Connecticut Society of Civil Engineers Annual Report* (1933): 21–53.

3. G. M. Turner and M. W. Jacobus, *Connecticut Railroads*, Hartford: Connecticut Historical Society, 1986, pp. 48–60.

4. Canfield & Robbins to the Housatonic Railroad Company, 17 March 1842. Horace Landon to the Housatonic Railroad Company, 28 December 1842. I thank Frederick Chesson for allowing me to see these letters. Receipts for the S. B. Moore & Co. shipments are at the Scoville Memorial Library in Salisbury.

5. Anon., *History of Litchfield County, Connecticut*, Philadelphia: Lewis, 1881, p. 101; Turner and Jacobus, pp. 88–98.

6. Robert B. Gordon, *American Iron, 1607–1900*, Baltimore: Johns Hopkins University Press, 1996, pp. 133–150, 244–246.

7. Ibid., p. 155.

8. Collins Company Historical Memoranda, Connecticut Historical Society ms. no. 72190, p. 300.

9. Gordon, *American Iron*, pp. 94–96, 129–132; Robert B. Gordon and

D. J. Killick, "The Metallurgy of the American Bloomery Process," *Archeomaterials* 6 (1992): 141–167.

10. Gregory Galer, Robert Gordon, and Frances Kemmish, *Connecticut's Ames Iron Works*, New Haven: Connecticut Academy of Arts and Sciences, 1998.

11. Francis Atwater, *History of Kent*, Connecticut, Meriden: Journal Publishing Company, 1897, p. 99.

12. Daniel P. Tyler, *Statistics of the Condition and Products of Certain Branches of Industry in Connecticut*, Hartford: Boswell, 1846.

13. Contract, Forbes & Adam papers, Connecticut Historical Society.

14. Forbes & Adam Washington Mills Ledger, 1795–1797; furnace accounts 1836–1838, both at the Falls Village–Canaan Historical Society.

15. Gordon, *American Iron*, pp. 112–114.

16. Holley & Coffing to Lee, 20 December 1820, Springfield Armory records.

17. William G. Neilson, *The Charcoal Blast Furnaces, Rolling Mills, Forges and Steel Works of New England in 1866*, Philadelphia: American Iron and Steel Association, 1866.

18. Contract, Forbes & Adam papers, Connecticut Historical Society.

19. Samuel Beckley's assertion (*Connecticut Western News*, 24 August 1899) that his father's Canaan blast furnace, built in 1846, was either the first or second in the United States with hot blast means he was misinformed about American developments elsewhere, unless he was referring only to the use of the Alger patent. Beckley should have known, since he grew up in a family of ironmasters and, according to his diary (Falls Village–Canaan Historical Society) had watched the North Adams furnace at work in 1863.

20. Zerah Colburn, *The Throttle Lever*, New York: Trow, 1856.

21. Ms. bills to the Housatonic Railroad from Canfield & Robbins. I thank Frederick Chesson for allowing me to see these documents.

22. Robert B. Gordon, "Material Evidence of Ironmaking Technique," *Journal of the Society for Industrial Archeology* 21 (1995): 69–80; Gordon, "Process Deduced from Ironmaking Wastes and Artifacts," *Journal of Archaeological Science* 24 (1997): 9–18; Gordon, "Materials For Manufacturing: The Response of the Connecticut Iron Industry to Limited Resources and Technological Change," *Technology and Culture* 24 (1983): 602–634. Felicia J. Deyrup, *Arms Makers of the Connecticut Valley*, Northampton, Mass.: Smith College, 1948, p. 136;

23. John Boyd, *Annals and Records of Winchester, Conn.*, Hartford: Case, Lockwood and Brainard, 1873, p. 510.

24. Galer, Gordon, and Kemmish, *Connecticut's Ames Iron Works; Connecticut Western News*, 6 October 1871.

25. Lawrence Van Alstyne, "Manufacturing in Sharon," Poconnuck Historical Society Collections, No. 1, Lakeville, 1912; Bruce Clouette, Sharon Valley National Register Nomination, 1979.

26. Robert B. Gordon and Michael S. Raber, "An Early American Integrated Steel Works," *Journal of the Society for Industrial Archeology* 10 (1984): 17–34.

27. Neilson, p. 231.

28. Galer, Gordon, and Kemmish, *Connecticut's Ames Iron Works*.

29. Horatio Ames to Oliver Ames, 5 April and 27 August 1844, Arnold B. Tofias Industrial Archives, Stonehill College.

30. Landon, Moore & Co. to Oliver Ames, 5 December 1843, Arnold B. Tofias Industrial Archives, Stonehill College.

31. Lee to Bomford, 1 August 1825, Springfield Armory records; ms. notes by Malcolm Rudd, Scoville Memorial Library, Salisbury.

32. Neilson, pp. 231, 240.

33. *Litchfield Enquirer*, 20 June 1867.

34. *Connecticut Western News*, 10 May 1872.

35. *Connecticut Western News*, 28 June 1872.

8. RETREAT FROM PROGRESS

1. Anon., *History of Litchfield County, Connecticut*, Philadelphia: Lewis, 1881; G. M. Turner and M. W. Jacobus, *Connecticut Railroads*, Hartford: Connecticut Historical Society, 1986; Robert W. Nimke, *The Central New England Railway Story*, vol. 3, Westmoreland, N. H.: Nimke, 1996. Bankruptcy and reorganizations after 1881 merged the Connecticut Western into the Central New England Railway, with tracks crossing the Hudson on the Poughkeepsie bridge. Nearly all service ended on the route in 1938.

2. A. L. Holley and L. Smith, "Salisbury Iron and the Works of the Barnum Richardson Company," *Engineering* 27 (1879): 451–452.

3. *Connecticut Western News*, 28 June 1872. At Ore Hill, the Mammoth Company had sixteen men and eight horses at work. The Chatfield Mining Company had fifty men and twenty horses raising 50 tons per day and expected to dig 15,000 tons in 1872. At the Davis mine, fifty-two men and thirty-one horses were raising 10 tons per day and expected to get 15,000 tons in 1872. Less than forty years earlier, the Davis mine was a small pit where a few men took ore out in baskets.

4. Data from the 1880 census. At Ore Hill, twenty of the miners now had to work underground as they penetrated to ore under deep overburden. They spent $3,669 for explosives and $430 for lumber in 1880.

5. These furnaces used 46,147 tons of ore, 7,573 tons of flux, and 2,134,162 bushels of charcoal valued at $222,992 to make 18,779 tons of pig iron, 684 tons of which were cold blast. One hundred ninety-nine men, one boy, ninety-eight horses and five steam engines with 105 horsepower raised 34,877 tons of ore (11,200 at Chatfield, 14,405 at Ore Hill, and 9,272 at Davis). Nearby mines in New York supplied the Connecticut furnaces with an additional 11,270 tons of ore. (Data from the 1880 U.S. Census.)

6. Connecticut Historical Society ms. nos. 74380 and 74381 have details.

7. Colliers' contracts with the Cornwall Bridge Iron Company through 1881 are in Connecticut Historical Society ms. no. 74153.

8. *Litchfield Enquirer*, 20 June 1867.

9. Emily M. Hopson, *Kent Tales: Iron Fever*, Kent, Conn.: Kent Historical Society, 1990, p. 47.

10. Kenneth D. LaFayette, *Flaming Brands: Fifty Years of Ironmaking in the Upper Peninsula of Michigan*, Marquette: Northern Michigan University Press, 1977, p. 13; Malcolm D. Rudd, *Men of Worth of Salisbury Birth*, Salisbury: Salisbury Association, 1991, p. 264.

11. Robert B. Gordon, *American Iron, 1607–1900*, Baltimore: Johns Hopkins University Press, 1996, p. 115. Unlike the older stoves set on the top of the furnace, the Player stove was placed at ground level, and was designed to attain complete combustion of the furnace gas.

12. James Bierce to Oliver Ames, 5 August 1874, 7 July 1875, Arnold B. Tofias Industrial Archives, Stonehill College. I thank Gregory Galer for finding these letters.

13. Alexander Lyman Holley to an unidentified cousin, letter no. 57 in the Salisbury Association collection.

14. "The Salisbury District," *Journal of the United States Association of Charcoal Iron Workers* 5 (1884): 361–368; "Narrative of the Sixth Annual Meeting," *Journal of the United States Association of Iron Workers* 6 (1885): 41–48. Continued reliance on waterpower for blowing engines in Salisbury while makers in other districts were turning to steam power is illustrated by the court action the Barnum-Richardson firm brought to prevent Thomas Stiles from diverting water from Mount Riga Brook, thereby diminishing the flow at the Lime Rock blast furnace. Ms. agreement between Barnum-Richardson and the heirs of T. A. Stiles, 2 June 1877. I thank Frederick Chesson for allowing me to see this document.

15. They were W. H. Barnum, C. W. Barnum, G. B. Burrall, N. A. McNeil, and M. B. Richardson.

16. A. L. Holley, "Notes on the Salisbury (Connecticut) Iron Mines and Works," *Transactions of the American Institute of Mining Engineers* 6 (1877–1878): 220–224; Holley and Smith, "Salisbury Iron and the Works of the Barnum Richardson Company," 451–452.

17. J. Lawrence Pool and Angeline J. Pool, *America's Valley Forges and Valley Furnaces*, West Cornwall: Pool, 1982, pp. 82–85; H. C. Keith, "The Early Iron Industry of Connecticut, Part I: History and Relics," *Connecticut Society of Civil Engineers Annual Report* 51 (1935): 148–175.

18. John Maloney, ms. at the Scoville Memorial Library, Salisbury.

19. J. F. Pynchon, "Iron Mining in Connecticut I, Ore Beds," "II, Smelting," "III, Historical Sketch," *Connecticut Magazine* 5 (1899): 20–26, 232–238, 277–285.

20. Anon., *Ironworks of the United States—1900*, Philadelphia: U.S. Iron and Steel Association, 1900.

21. J. M. Camp and C. B. Francis, *The Making, Shaping and Treating of Steel*, 4th ed., Pittsburgh: Carnegie Steel Co., 1925, p. 620.

22. Lawrence Eddy, "The Old Ironworks of Canaan and Vicinity," ms. at the Falls Village–Canaan Historical Society, 1969; *Connecticut Western News*, 29 August 1935. (Lawrence Eddy worked at the retort plant in 1919.)

23. Malcolm Rudd interview with Charles F. Perkins, 27 September 1935, ms. at the Scoville Memorial Library, Salisbury. Perkins also said that Barnum later had poor advisers and was gypped by them. Barnum was a good salesman, but many of his orders "had to be scrutinized from the point of view of profits, etc."

24. J. E. Johnson Jr., "Report on the Barnum-Richardson Company," 1916, ms., Scoville Memorial Library, Salisbury.

25. Perkins's comment in interview with Malcolm Rudd, ms., Scoville Memorial Library, Salisbury.

26. Anon., *Ironworks of the United States—1900*, Eddy mss., 1969; Kenneth Howell and Einar W. Carlson, *Men of Iron: Forbes and Adam*, Lakeville, Conn.: Pocketknife Press, 1980, pp. 121–122.

27. Bankruptcy papers, Salisbury Iron Corporation, Scoville Memorial Library, Salisbury.

28. Annual Report of the Salisbury Iron Corporation; R. Moldenke, *Charcoal Iron*, Lime Rock, Conn.: Salisbury Iron Corporation, 1920.

1. Data for the dates of construction and abandonment of forges and furnaces define the number and type of ironworks extant at any given time (Appendix 1). The data for blast furnaces are complete, and those for the finery forges and puddling works, nearly so. The number of extant works year by year are shown in figures 9.1 and 9.2. We have only a poor record of the dates of the construction and, particularly, of abandonment for the bloomery forges. The cumulative number of bloomeries built based on available dates is shown in figure 9.3.

2. Zerah Colburn, *The Throttle Lever*, New York: Trow, 1856.

3. Robert B. Gordon, "Materials for Manufacturing: The Response of the Connecticut Iron Industry to Limited Resources and Technological Change," *Technology and Culture* 24 (1983): 602–634. Lack of understanding of the relation of ore composition and the properties of the metal produced from it is shown by the belief held by artisans in Winsted that the Salisbury iron was deficient in an ingredient (unknown to them) necessary for its conversion to steel of edge-tool quality. (Anon., *History of Litchfield County, Connecticut*, Philadelphia: Lewis, 1881, p. 201).

4. "Amount of Dividends Made on Ore Sold," ms. Coffing Estate Box, Connecticut Historical Society.

5. *Connecticut Western News*, 24 November 1871.

6. *Connecticut Western News*, 20 October 1871.

7. *Connecticut Western News*, 6 October 1871.

8. Wilson to Landon, Moore & Co., 5 July 1839, Barnum-Richardson papers, University of Connecticut, Storrs.

9. M. D. Rudd, "Lakeville in the American Switzerland," *Connecticut Magazine* 8 (1903): 337–371, see p. 369; *Gleason's Pictorial Drawing Room Companion* 6, no. 23 (10 June 1854): 360–361.

10. Henry W. Beecher, *Star Papers or Experiences of Art and Nature*, New York: Derby, 1855, pp. 137–157.

11. Benjamin Silliman, "Sketches of a Tour in the Counties of New Haven and Litchfield in Connecticut," *American Journal of Science* 2 (1820): 201–235; see p. 215.

12. Howard S. Russell, *A Long Deep Furrow: Three Centuries of Farming in New England*, ed. Mark Lapping, Hanover, N.H.: University Press of New England, 1982, p. 132; Edward C. Kirkland, *Men, Cities and Transportation, 1820–1900*, Cambridge, Mass.: Harvard University Press, 1948, p. 308; George M. Milne, *Connecticut Woodlands*, Connecticut Forest and Park Commission, 1995, pp. 2, 15. Connecticut finally required railroads to install required spark arresters on their locomotives in 1923.

13. These figures are computed from the output and fuel ratio of the blast furnaces.

14. Interview with Malcolm Rudd, 27 September 1935; ms. notes at the Scoville Memorial Library, Salisbury.

15. *Lakeville Journal*, 27 July 1961.

16. *Connecticut Western News*, 29 August 1935.

17. Malcolm D. Rudd, *Men of Worth of Salisbury Birth*, Salisbury: Salisbury Association, 1991, p. 146.

18. Marcia Holley Rudd, *Lakeville Journal*, 7 January 1907.

19. D. C. Kilbourn, "Washington," *Connecticut Quarterly* 4 (1898): 237–256.

20. Milne, pp. 23, 81.

21. Chard P. Smith, *The Housatonic,* New York: Rinehart, 1946, pp. 386, 394.

22. Ibid., p. 418. E. B. Eaton, "Lakeville—Its Educational and Commercial Interests," *Connecticut Magazine* 8 (1903): 372–384.

23. National Register nomination, Lime Rock Historic District.

24. Iveagh Hunt Sterry and William H. Garrigus, *They Found a Way,* Brattleboro, Vt.: Stephen Day Press, 1938, pp. 106–119.

25. Smith, *The Housatonic,* p. 377

10. COMMUNITY, CULTURE, AND INDUSTRIAL ECOLOGY

1. Exhaustion of ore was never an issue in Salisbury. More ore would have been taken from the mine at Ore Hill during World War II had not the railroad that served it been abandoned several years earlier.

2. See, for example, Robert B. Gordon, "Custom and Consequence: Early Nineteenth-Century Origins of the Environmental and Social Costs of Mining Anthracite," in *Early American Technology,* ed. J. McGaw, Chapel Hill: University of North Carolina Press, 1994, pp. 240–277.

3. Charles B. Dew, *Ironmaker to the Confederacy: Joseph R. Anderson and the Tredegar Iron Works,* New Haven, Conn.: Yale University Press, 1966; Dew, *Band of Iron, Master and Slave at Buffalo Forge,* New York: Norton, 1994.

4. Anne K. Knowles, *Calvinists Incorporated,* Chicago: University of Chicago Press, 1997, pp. 177–224.

APPENDIX 1. IRONWORKS INVENTORY

1. Charles R. Harte, "The Early Connecticut Iron Industry, Part II: The Connecticut Blast Furnaces and Furnace Practice," *Connecticut Society of Civil Engineers Annual Report* 51 (1935): 176–214; H. C. Keith, "The Early Iron Industry of Connecticut, Part I: History and Relics," *Connecticut Society of Civil Engineers Annual Report* 51 (1935): 148–175; Edward Kirby, *Echoes of Iron,* Sharon, Conn.: Sharon Historical Society, 1998.

2. Gregory Galer, Robert Gordon, and Frances Kemmish, *Connecticut's Ames Iron Works,* New Haven: Connecticut Academy of Arts and Sciences, 1998.

3. Locations and descriptions of the sites of many of the ironworks discussed here are found in Robert B. Gordon and Michael S. Raber, *Industrial Heritage in Northwestern Connecticut,* New Haven: Connecticut Academy of Arts and Sciences, 2000.

APPENDIX 2. IRONMAKING AND STEELMAKING TECHNIQUES

1. Robert B. Gordon, *American Iron, 1607–1900,* Baltimore: Johns Hopkins University Press, 1996.

2. Robert B. Gordon, "Material Evidence of Ironmaking Technique," *Journal of the Society for Industrial Archeology* 21 (1995): 69–80.

3. Charles R. Harte, "The Early Connecticut Iron Industry, Part II: The Connecticut Blast Furnaces and Furnace Practice," *Connecticut Society of Civil Engineers Annual Report* 51 (1935): 176–214; Harte, "Connecticut's

Cannon," *Connecticut Society of Civil Engineers Annual Report* 58 (1942): 52–70.

4. James A. Mulholland, *A History of Metals in Colonial America*, Tuscaloosa: University of Alabama Press, 1981, p. 23. There has been no archaeological study of the East Haven (ca.1660) or Lakeville (1762) furnaces. The often-reproduced drawing of the Lakeville furnace was first published in late nineteenth-century promotional literature and shows a furnace of about the same size as that on Mount Riga but with three blowing arches.

5. Excavations at Valley Forge in Pennsylvania show that the artisans there used a forge similar to the Walloon fineries common in England in the previous century.

6. Robert B. Gordon, "Choice of Method for Making Wrought Iron in the Salisbury District of Connecticut," *Journal of the Historical Metallurgy Society* 31, no. 1 (1997): 25–31.

7. Gordon, *American Iron*, pp. 99, 116, 133.

APPENDIX 3. UNITS AND CONVERSION FACTORS

1. These factors are based on data reported by Herbert H. Winer, "History of the Great Mountain Forest, Litchfield County, Connecticut," Ph.D. diss., Yale University, 1955.

Bibliography

Allen, Richard S., "Connecticut Iron and Steel from Black Sea Sands," *IA, Journal of the Society for Industrial Archeology* 18 (1992): 129–132.

Altamura, Robert J., *Bedrock Mines and Quarries of Connecticut*, 1987 (map and citation list).

Alvord, Eliphaz, *Winchester-Winsted in 1813*, Hartford: Acorn, 1961.

Anon., *History of Litchfield County, Connecticut*, Philadelphia: Lewis, 1881.

Anon., *Ironworks of the United States—1900*. Philadelphia: U.S. Iron and Steel Association, 1900.

Anon., "Salisbury Iron, Its Composition, Qualities, and Uses," reprinted from the *New York Railroad Gazette*, Salisbury: Barnum, Richardson Co., 1878.

Atwater, Francis, *History of Kent, Connecticut*, Meriden: Journal Publishing Company, 1897.

Barber, John W., *Connecticut Historical Collections*, New Haven: Durrie, Pack and Barber, 1838.

Beecher, Henry W., *Star Papers or Experiences of Art and Nature*, New York: Derby, 1855.

Bidwell, Percy W., "Rural Economy in New England at the Beginning of the Nineteenth Century," *Transactions of the Connecticut Academy of Arts and Sciences* 20 (1916): 241–399.

Blodget, William, *A New and Correct Map of Connecticut*, Middletown, 1792.

Boyd, John, *Annals and Records of Winchester, Conn.*, Hartford: Case, Lockwood and Brainard, 1873.

Camp, J. M., and C. B. Francis, *The Making, Shaping and Treating of Steel*, 4th ed., Pittsburgh: Carnegie Steel Co., 1925.

Church, Samuel, "Historical Address" (delivered at the Centennial Town Meeting of Salisbury, 20 October 1841), in *Historical Addresses*, Pittsfield, 1876.

Clark, Christopher, *The Roots of Rural Capitalism*, Ithaca, N.Y.: Cornell University Press, 1990.

Clark's Map of Litchfield County, surveyed by G. M. Hopkins Jr., Philadelphia: Richard Clark, 1859.

Colburn, Zerah, *The Throttle Lever*, New York: Trow, 1856.

Crossman, David W., "A New Year's Discourse Delivered at Salisbury . . . January 2nd 1803: Containing the Ancient History of the Town," Hartford: Hudson and Goodwin, 1803.

Daggett, David, "A Brief Account of a Trial at Law . . . ," *Memoirs of the Connecticut Academy of Arts and Sciences* 1, no. 1 (1810): 131–134.

Dana, J. D., "Berkshire Beds of Iron," *Berkshire Historical and Scientific Society Collections* 1 (1886): 19–25.

Daniels, Bruce C., *The Connecticut Town*, Middletown: Wesleyan University Press, 1979.

Dew, Charles B., *Ironmaker to the Confederacy: Joseph R. Anderson and the Tredegar Iron Works*, New Haven, Conn.: Yale University Press, 1966.

———, *Band of Iron: Master and Slave at Buffalo Forge*, New York: Norton, 1994.

Deyrup, Felicia J., *Arms Makers of the Connecticut Valley*, Northampton, Mass.: Smith College, 1948.

Dwight, Timothy, *Travels in New England and New York*, 4 vols., New Haven: T. Dwight, 1821.

Eaton, E. B., "Lakeville—Its Educational and Commercial Interests," *Connecticut Magazine* 8 (1903): 372–384.

Eddy, Lawrence, "The Old Ironworks of Canaan and Vicinity," ms. at the Falls Village–Canaan Historical Society, 1969.

Ekern, E. A. "The Falls Village Hydroelectric Development of the Connecticut Power Co.," *Connecticut Society of Civil Engineers Annual Report* (1913): 110–130.

Galer, Gregory, Robert Gordon, and Frances Kemmish, *Connecticut's Ames Iron Works*, New Haven: Connecticut Academy of Arts and Sciences, 1998.

Gates, R. M., *The Bedrock Geology of the Sharon Quadrangle*, State Geological and Natural History Survey of Connecticut, Quadrangle Report No. 38, 1979.

Geographical Location of Historical Landmarks in the Town of Salisbury, Connecticut, Salisbury: Salisbury Association, 1949.

Gordon, Robert B., "English Iron for American Arms: Laboratory Evidence on the Source of Iron Used at the Springfield Armory in 1860," *Journal of the Historical Metallurgy Society* 17 (1983): 91–98.

———, "Materials for Manufacturing: The Response of the Connecticut Iron Industry to Limited Resources and Technological Change," *Technology and Culture* 24 (1983): 602–634.

———, "Hydrological Science and the Development of Water Power for Manufacturing," *Technology and Culture* 26 (1985): 204–235.

———, "Strength and Structure of Wrought Iron," *Archeomaterials* 2 (1988): 109–137.

———, "Custom and Consequence: Early Nineteenth-Century Origins of the Environmental and Social Costs of Mining Anthracite," in *Early American Technology*, ed. J. McGaw, pp. 240–277, Chapel Hill: University of North Carolina Press, 1994.

———, "Material Evidence of Ironmaking Technique," *Journal of the Society for Industrial Archeology* 21 (1995): 69–80.

———, *American Iron, 1607–1900*, Baltimore: Johns Hopkins University Press, 1996.

———, "Choice of Method for Making Wrought Iron in the Salisbury District of Connecticut," *Journal of the Historical Metallurgy Society* 31, no. 1 (1997): 25–31.

———, "Process Deduced from Ironmaking Wastes and Artifacts," *Journal of Archaeological Science* 24 (1997): 9–18.

———, "Travel on Connecticut's Roads, Bridges, and Ferries, 1790–1830," in *Connecticut Towns, 1800–1832: Voices of the New Republic*, vol. 2, *What We Think*, ed. Howard Lamar and Carolyn Cooper, New Haven: Connecticut Academy of Arts and Sciences, 2000.

Gordon, Robert B., and D. J. Killick, "The Metallurgy of the American Bloomery Process," *Archeomaterials* 6 (1992): 141–167.

Gordon, Robert B., and Patrick M. Malone, *The Texture of Industry*, New York: Oxford University Press, 1994.

Gordon, Robert B., and Michael S. Raber, "An Early American Integrated Steel Works," *Journal of the Society for Industrial Archeology* 10 (1984): 17–34.

———, *Industrial Heritage in Northwestern Connecticut, a Guide to History and Archaeology*, New Haven: Connecticut Academy of Arts and Sciences, 2000.

Gordon, Robert B., and Geoffrey Tweedale, "Pioneering in Steelmaking at the Collins Axe Factory," *Journal of the Historical Metallurgy Society* 24 (1990): 1–11.

Graedel, T. E., and B. R. Allenby, *Industrial Ecology*, Englewood Cliffs, N.J.: Prentice Hall, 1995.

Grant, Charles S., *Democracy in the Connecticut Frontier Town of Kent*, New York: Columbia University Press, 1961.

Harris, H. P., "Charcoal Manufacture in the Salisbury Iron Region," *Journal of the United States Association of Charcoal Ironworkers* 6 (1885): 49–53.

Harte, Charles R., "Connecticut's Canals," *Connecticut Society of Civil Engineers Annual Report* (1933): 21–53.

———, "The Early Connecticut Iron Industry, Part II: The Connecticut Blast Furnaces and Furnace Practice," *Connecticut Society of Civil Engineers Annual Report* 51 (1935): 176–214.

———, "Connecticut's Cannon," *Connecticut Society of Civil Engineers Annual Report* 58 (1942): 52–70.

Hobbs, William H., "The Iron Ores of the Salisbury District of Connecticut and Massachusetts," *Economic Geology* 2 (1907): 153–181.

Holley, A. L., "Notes on the Salisbury (Connecticut) Iron Mines and Works," *Transactions of the American Institute of Mining Engineers* 6 (1877–1878): 220–224.

Holley, A. L., and L. Smith, "Salisbury Iron and the Works of the Barnum Richardson Company," *Engineering* 27 (1879): 451–452.

Hopson, Emily M., *Kent Tales: Rufus Fuller and the South Kent Ore Bed*, Kent, Conn.: Kent Historical Society, 1988.

———, *Kent Tales: Iron Fever*, Kent, Conn.: Kent Historical Society, 1990.

Howell, Kenneth, and Einar W. Carlson, *Empire over the Dam*, Chester, Conn.: Pequot Press, 1974.

———, *Men of Iron: Forbes and Adam*, Lakeville, Conn.: Pocketknife Press, 1980.

Johnson, J. E., Jr., "Report on the Barnum-Richardson Company," ms., Scoville Memorial Library, Salisbury, 1916.

Keith, H. C., "The Early Iron Industry of Connecticut, Part I: History and Relics," *Connecticut Society of Civil Engineers Annual Report* 51 (1935): 148–175.

Kilbourn, D. C., "Washington," *Connecticut Quarterly* 4 (1898): 237–256.

Kirby, Edward, *Echoes of Iron*, Sharon, Conn.: Sharon Historical Society, 1998.

Kirkland, Edward C., *Men, Cities and Transportation, 1820–1900*, Cambridge, Mass.: Harvard University Press, 1948.

Knowles, Anne K., *Calvinists Incorporated*, Chicago: University of Chicago Press, 1997.

LaFayette, Kenneth D., *Flaming Brands: Fifty Years of Ironmaking in the Upper Peninsula of Michigan*, Marquette: Northern Michigan University Press, 1977.

Main, Jackson T., *Society and Economy in Colonial Connecticut*, Princeton, N.J.: Princeton University Press, 1985.

Manchester, Irving E., *The History of Colebrook*, Colebrook, Conn.: Sesquicentennial Committee, 1935.

McGaw, Judith, *Most Wonderful Machine: Mechanization and Social Change in Berkshire Paper Making, 1801–1885*, Princeton, N.J.: Princeton University Press, 1987.

McHugh, Jeanne, *Alexander Holley and the Makers of Steel*, Baltimore: Johns Hopkins University Press, 1980.

McManis, Douglas R., *Colonial New England: A Historical Geography*, New York: Oxford University Press, 1975.

Merchant, Carolyn, *Ecological Revolutions, Nature, Gender, and Science in New England*, Chapel Hill: University of North Carolina Press, 1989.

Middlebrook, Louis F., *Connecticut's Cannon*, Salem, Mass.: Newcomb Gauss, 1935.

Miller, Harry, "Potash from Woodashes: Frontier Technology in Canada and the United States," *Technology and Culture* 21 (1980): 187–208.

Milne, George M., *Connecticut Woodlands*, Connecticut Forest and Park Commission, 1995.

"Mining Industries of the U.S.," 10th Census, vol. 15 (1880): 134. (Description of the Salisbury Mines.)

Mitchell, Isabel S., *Roads and Road-making in Colonial Connecticut*, New Haven, Conn.: Yale University Press, 1933.

Moldenke, R., *Charcoal Iron*, Lime Rock, Conn.: Salisbury Iron Corporation, 1920.

Morrill, William F., "The Spur at the Head of the Manor, Salisbury, Connecticut, and the Livingstons," in *The Livingston Legacy*, ed. Richard T. Wiles, pp. 38–66, Annandale, N.Y.: Bard College, 1987.

Morris, James, *Statistical Account of Several Towns in the County of Litchfield*, pp. 85–124 of *A Statistical Account of the Towns and Parishes of Connecticut*, New Haven: Connecticut Academy of Arts and Sciences, 1811.

Morse, Jarvis M., *A Neglected Period in Connecticut's History, 1818–1850*, New Haven, Conn.: Yale University Press, 1933; reprint, New York: Octagon Books, 1978.

Mulholland, James A., *A History of Metals in Colonial America*, Tuscaloosa: University of Alabama Press, 1981.

"Narrative of the Sixth Annual Meeting," *Journal of the United States Association of Charcoal Iron Workers* 6 (1885): 41–48

Neilson, William G., *The Charcoal Blast Furnaces, Rolling Mills, Forges and Steel Works of New England in 1866*, Philadelphia: American Iron and Steel Association, 1866.

Newland, D. H., "Mineralogy and Origin of the Taconic Limonites," *Economic Geology* 31 (1936): 133–155.

Nimke, Robert W., *The Central New England Railway Story*, vol. 3, Westmoreland, N.H.: Nimke, 1996.

Norton, Lewis M., "Goshen," in *Connecticut Towns, 1800–1832: Voices of the New Republic*, vol. 1, *What They Said*, ed. Howard Lamar and Carolyn Cooper, New Haven: Connecticut Academy of Arts and Sciences, 2000.

Orcutt, Samuel, *History of the Towns of New Milford and Bridgewater*, Hartford: Case, Lockwood and Brainard, 1882.

Pease, John C., and John M. Niles, *A Gazetteer of the States of Connecticut and Rhode Island*, Hartford: Marsh, 1819.

Pettee, Julia, *The Rev. Jonathan Lee and His Eighteenth Century Parish: The Early History of the Town of Salisbury, Connecticut*, Salisbury: Salisbury Association, 1957.

Pool, J. Lawrence, and Angeline J. Pool, *America's Valley Forges and Valley Furnaces*, West Cornwall: Pool, 1982.

Pownall, T., *A Topographical Description of the Dominion of the United States of America*, Pittsburgh: University of Pittsburgh Press, 1949.

Prude, Jonathan, "Town-Factory Conflicts in Antebellum Rural Massachusetts," in *The Countryside in the Age of Capitalist Transformations*, ed. S. Hahn and J. Prude, Chapel Hill: University of North Carolina Press, 1985.

Purcell, Richard J., *Connecticut in Transition*, 1918; republished, Middletown: Wesleyan Uniiversity Press, 1963.

Pynchon, J. F., "Iron Mining in Connecticut I, Ore Beds," "II, Smelting," "III Historical Sketch," *Connecticut Magazine* 5 (1899): 20–26; 232–238, 277–285.

Rand, Christopher, *The Changing Landscape*, New York: Oxford University Press, 1968.

Rockwell, Francis W., *The Rockwell Family*, Pittsfield, Mass.: privately published, 1924.

Rogers, Winton B., "Short History of the Water Power Development at Falls Village," *The Lure of the Litchfield Hills* 8 (1951): 301, 315, 332.

Rudd, Malcolm E., "Lakeville in the American Switzerland," *Connecticut Magazine* 8 (1903): 337–371.

———, *Men of Worth of Salisbury Birth*, Salisbury: Salisbury Association, 1991.

Russell, Howard S., *A Long Deep Furrow: Three Centuries of Farming in New England*, ed. Mark Lapping, Hanover, N.H.: University Press of New England, 1982.

"The Salisbury District," *Journal of the United States Association of Charcoal Iron Workers* 5 (1884): 361–368.

"Salisbury Iron, Its Composition, Qualities and Uses," in *The New York Railway Gazette*, reprinted, Barnum Richardson Co., 1878.

Salisbury Town Meeting Minutes, 1741–1784, ed. Geoffrey Rossano, Salisbury: Salisbury Association, 1988.

Schairer, J. F., *The Minerals of Connecticut*, State Geological and Natural History Survey of Connecticut, Bulletin 51, 1931.

Shepard, Charles U., *A Report on the Geological Survey of Connecticut*, New Haven, 1837.

Silliman, Benjamin. "Sketches of a Tour in the Counties of New Haven and Litchfield in Connecticut," *American Journal of Science* 2 (1820): 201–235.

Skinner, B. J., "A Second Iron Age Ahead," *American Scientist* 64 (1976): 258–269.

———, "Earth Resources," *Proceedings of the National Academy of Sciences* 76 (1979): 4212–4217.

Slosson, Barzillai, "Kent," in *Connecticut Towns, 1800–1832: Voices of the New Republic*, vol. 1, *What They Said*, ed. Howard Lamar and Carolyn Cooper, New Haven: Connecticut Academy of Arts and Sciences, 2000.

Smith, Chard P., *The Housatonic*, New York: Rinehart, 1946.

Smith, John C., "Sharon," in *Connecticut Towns, 1800–1832: Voices of the New Republic*, vol. 1, *What They Said*, ed. Howard Lamar and Carolyn Cooper, New Haven: Connecticut Academy of Arts and Sciences, 2000.

Smith, M. R., *Harpers Ferry Armory and the New Technology*, Ithaca, N.Y.: Cornell University Press, 1977.

Sterry, Iveagh Hunt, and William H. Garrigus, *They Found a Way*, Brattleboro, Vt.: Stephen Day Press, 1938.

Taylor, Philip E., "The Turnpike Era in New England," Ph.D. diss., Yale University, 1934

Thorson, Robert M., and others, "Colonial Impacts to Wetlands in Lebanon, Connecticut," chap. 3 in *Reviews in Engineering Geology*, vol. 12, ed. C. W. Webb and M. E. Gowan, Boulder, Colo.: Geological Society of America, 1998.

Topographical and Historical Map of the Town and Villages of Salisbury, topography by I. W. and E. F. Sanford, historical sketches by Malcolm D. Rudd and I. W. Sanford, New York: Sanford, 1899.

Turner, G. M., and M. W. Jacobus, *Connecticut Railroads*, Hartford: Connecticut Historical Society, 1986.

Tyler, Daniel P., *Statistics of the Condition and Products of Certain Branches of Industry in Connecticut*, Hartford: Boswell, 1846.

Van Alstyne, Lawrence, "Manufacturing in Sharon," Poconnuck Historical Society Collections, No. 1, Lakeville, 1912.

Warren, Moses, and William Gillet, *Connecticut from Actual Survey Made 1811*, Hartford: 1812.

Warren, William L., "The Litchfield Iron Works in Bantam, Connecticut," *Tools and Technology*, July 1992, pp. 1–27.

Winer, Herbert H., "History of the Great Mountain Forest, Litchfield County, Connecticut," Ph.D. diss., Yale University, 1955.

Wood, Fredric J., *The Turnpikes of New England*, Boston: Jones, 1919.

Zen, E., and J. H. Hartschorn, "Geologic Map of the Bashbish Quadrangle," Map GQ-507, Washington, D.C.: U.S. Geological Survey, 1966.

Index

Absentee owners, lack of, 111;
W. H. Barnum as, 114
Account books, kept by artisans, 27
Adam, Charles S., 74
Adam, Forbes S., 74
Adam, George, 74
Adam, John Jr., 41, 61; library, 64;
marries Abigail Forbes, 40;
partner with Samuel Forbes,
40
Adam, John III, 42
Adam, Leonard, 41
Adam, Samuel Forbes, 41, 65
Adam, William, 42
Adirondack ironmaking compared to
Salisbury, 117
Agricultural products, important in
district's economy, 63–64
Allen, Ethan, 25, 28; failure as a
manager, 30
American bloomery process, 27, 71
American Institute of Mining
Engineers, visit to Salisbury, 88
Ames, Horatio, 60–61, 71, 113;
develops rolling machinery, 81; a
farmer, 63; visits ironworks in
other districts, 62
Ames ironworks, 72–73; 76, 78–80, 97,
99
Ames, Oliver, buys Salisbury iron, 52
Amesville, 72
Anchors, for USS *Franklin*, 43; guaran-
teed, 39
Animals, draft, 86
Antietam forge, fails to supply Harpers
Ferry Armory, 117

Armories, as Salisbury customers, 35.
See also Harpers Ferry Armory,
Pomeroy, Lemuel, Springfield Armory, Whitney, Eli
Artisans, independence of, 60; as inside
contractors, 41; opportunities for
advancement, 35
Artisans' skills, in furnace construction, 38; managerial, 60; opportunities for learning, 60; reliance
on, 36, 44, 113; taxed by demand
for uniformity, 27, 53
Artisans' work, pride taken in, 60; risks
in, 60; unlike that in manufacturing, 60
Ashley, Ezekiel, 12
Austin, Benjamin, 23
Austin, Thomas, 23

Bachelor's Club, 64
Bacon, Jacob, 22
Bainbridge, Commodore William, 35
Barnum, C. W., 89
Barnum, Ebenezer, 22
Barnum, Milo, 63, 74, 83
Barnum, R. N., 89
Barnum, W. M., 92
Barnum, William H., 63, 66, 71, 83
Barnum-Richardson Company, 83–85;
bankrupt, 93; creditors receive
stock Salisbury Iron Corporation,
93; Johnson's report on, 92–93;
managerial stagnation of, 87, 93;
vertical integration, 84
Beckley, John, 74
Bedrock, map, 16

Spencer, William, 24
Springfield Armory, 51; Alexander Holley visits, 47; demand for gun iron in Civil War, 81; institutes gun iron tests, 51; purchases iron from Canfield & Robbins, 52, from Salisbury Center Forge, 50; purchases Salisbury iron, 44; railroad deliveries of iron, 70
Standardization of products, 39–40
Steam hammers, adopted at Ames ironworks, 78
Steel, difficulties with quality, 45; made by Rockwells, 44; making described, 45
Steel puddling works at Mine Hill, 80
Steel wheels, adopted by railroads, 92
Sterling, Chapin & Company, 62
Sterling, Elisha, 36
Sterling, F. A., 62
Sterling, Landon & Company, 62
Sterling, W. C., 37, 62, 69
Stoddard, Josiah, 24
Stone, Alfred, 105
Sustainability, defined, 3

Technology, challenges from new, 68; new, spurned, 67, 87, 71, transfer of, 61, 87; use of obsolete, 87, 92
Temperance movement, 49, 113
Tourists, attracted to district, 99, 115
Turnpike roads, 16–17; map, 18

Uniformity, demand for in iron, 27; needed for manufacturing, 51; problems in attaining, 35
U.S. Association of Charcoal Iron Workers, visit to Salisbury, 88–89

Verrill, A. E., 98

Wadsworth, Decius, 36
Wadsworth & Kirby, 27, 61
Wages, paid at Lakeville blast furnace, 34
Walker, Robert Jr., 13, 21, 22–23

War, effect on demand, 25; effect on prices, 37; ordnance contracts for, 80
Washington, D.C., described, 48, 64
Water Power Company, 66, 82
Water privileges, 13; agreements for mutual use, 23, 31, 57; full utilization of, 41; Highland Lake, 31
Waterman, David, 36
Waterpower systems, environmental effects of, 46, 57; management of, 57
Weed, Hiram, 62
West Cornwall, 70
Western Lands, 7; iron ore in, 11; natural resources in, 11; origin of, 12; towns formed in, 12
Western Railroad, 50, 68
Wheels, railroad car, 74
Whitcomb, Edward, 33
White, Zebulon, 33
Whiting, Joseph, 33
Whitney, Eli, 43
Williams, Benajah, 24
Williams, Elisha, 13, 21
Williams, Phineas, 45
Wolcott, Alexander, 13
Wononscopomuc Lake, 15, 17, 28
Wooden, Abner and Peter, 36
Woodland, burning by Indians, 13; changes in area, 101; considered inexhaustible, 100; consumption rate, 54; coppice, 14; demands on, 100–101; managed for sustained production, 14, 56–57, 111; purchased by ironmakers, 56; purchased for conservation and recreation, 104, 107; requirements, 56
Woodville, 25–26; map of, 26; slitting mill built, 41
Workers' houses, 73
Wrought iron, demand for, 20, 98, 114; production abandoned, 71. *See also* gun iron

Printed in the United States
21162LVS00001B/89